SPEED READ
SUPERCAR

Inspiring | Educating | Creating | Entertaining

Brimming with creative inspiration, how-to projects, and useful information to enrich your everyday life, Quarto Knows is a favorite destination for those pursuing their interests and passions. Visit our site and dig deeper with our books into your area of interest: Quarto Creates, Quarto Cooks, Quarto Homes, Quarto Lives, Quarto Drives, Quarto Explores, Quarto Gifts, or Quarto Kids.

10 9 8 7 6 5 4 3 2 1

ISBN: 978-0-7603-6291-4

Digital edition published in 2018
eISBN: 978-0-7603-6292-1

Library of Congress Control Number: 2018945243

Acquiring Editor: Zack Miller
Project Manager: Jordan Wiklund
Series Creative Director: Laura Drew
Cover and interior illustrations by Rafael Santos de Oliveira

Printed in China

MIX
Paper from
responsible sources
FSC® C017606

SPEED READ
SUPERCAR

THE HISTORY, TECHNOLOGY AND DESIGN
BEHIND THE WORLD'S MOST EXCITING CARS

BASEM WASEF

INTRODUCTION

What makes a supercar *super*? The term is admittedly hyperbolic and liberally bandied about, but what truly warrants the weighty moniker? Is it the sweep of a silhouette? A top speed above 200 mph? Upward sweeping doors, an astronomic price tag, or maybe an excessive cylinder count? Arriving at a definition is not easy and has triggered countless debates.

These pages attempt to decode the big-picture concepts, groundbreaking engineering, and imaginative designs that elevate supercars beyond more mundane forms of four-wheeled transportation. But a supercar is more than just its physical attributes; we must also consider the historical trends that define the genre, the tumultuous personalities behind these passion-driven projects, and the game-changing technology, much of which trickles down from the racing world, that enables such extraordinary performance.

Historically, the term *supercar* emerged in 1920 in a newspaper advertisement for a sleek, big-engined cabriolet. But it didn't enter the collective consciousness until decades later. Groundbreaking sports cars such as the 1954 Mercedes-Benz 300SL captured the imagination of enthusiasts everywhere with its shockingly cool gullwing doors and state-of-the-art, race-derived hardware, and in 1961 the Jaguar E-Type wowed the world with its slinky lines and lithesome athleticism, inspiring none other than Enzo Ferrari to decree it "The most beautiful car ever made." But the most widely acknowledged instigator of the supercar trend was the Lamborghini Miura, a low-slung, V-12-powered curlicue of a car that debuted in 1966 and inspired enough shock and awe to earn it supreme status in the supercar pantheon. The Miura's impact was so significant that Ferrari was forced to retaliate with the stirring Daytona.

In a way, the seismic effect of the Lamborghini Miura captures the elusive essence of what makes a supercar. While it claimed gorgeous bodywork and a symphonious V-12 engine, its doors opened conventionally, and it famously couldn't breach the mythical 200-mph threshold. Despite its incomplete matrix of qualifications, though, the Miura is widely considered the world's first supercar.

If anything, the idea of a supercar becomes nearly irrelevant because the word carries so much subjective emotional weight that it cannot be captured with one sweeping definition. It has even inspired "hypercar," a spinoff term that attempts to express an even more superlative type of high-performance vehicle. I, for one, liken supercars to pornography: though pundits debate what it really means, you know it when you see it.

THE GROUNDBREAKERS

THE GROUNDBREAKERS
MERCEDES-BENZ 300 SL "GULLWING"

FUN FACT

The 300 SL suffered from awkward entry and exit due to the body's thick door sills and a notoriously hot cabin thanks to copious glass and fixed windows.

HISTORICAL TIDBIT

With only 1,400 300 SL coupes produced between 1954 and 1957, the model commands seven-figure values. Rarer still are the alloy-body models, of which only twenty-nine were built. Intended for competition and certified for racing, these special 300 SLs shaved off around 200 pounds .

KEY PERSON

Chief Engineer Rudolf Uhlenhaut who made his inimitable mark on the car thanks to his background as a hands-on racing engineer, supervised legendary wins such as Stirling Moss's record-setting 1955 Mille Miglia victory.

More than a decade before the wild Lamborghini Miura earned unofficial designation as the world's first supercar, Mercedes-Benz pulled the curtain off the 300 SL, an elegant and outrageously capable machine that some argue was equally worthy of the superlative. While Mercedes road cars conventionally focused on stodgy German priorities such as engineering and safety, the 300 SL used the brand's racing experience to bring an extra shot of performance to the road. The resulting two-seater offered a level of exoticism previously unseen in a production Mercedes, elevating the brand with a moonshot of functional and emotional panache.

Key to the 300 SL's identity are its trademark gullwing doors, which were developed as a packaging necessity, not a styling gimmick. The trick setup was employed because the car's lightweight chassis, consisting of a complex lattice of alloy tubes, created unusually high side sills. Since a tiny door opening would have made cabin access all but impossible for normal-sized adults, the upwardly sweeping apertures offered a solution that was reasonably practical and entirely sexy.

The Gullwing's curvaceous body is unequivocally attractive, but virtually every aspect of its form is derived from function. For instance, the gentle "eyebrow" arches over the wheels aid aerodynamics, while the engine is canted 50 degrees on its side to facilitate a low center of gravity and a sleek hood line. The 3-liter, inline six-cylinder powerplant was the first to incorporate an advanced direct fuel injection system, enabling an output of 212 horsepower—nearly double that of the carbureted race car from which it was derived.

Though tame by modern standards, the 300 SL had a combination of relatively high power and lightweight construction for the midcentury era that made it the fastest production car of its day. Costing over $7,000 new—a small fortune for the mid-1950s—the Gullwing's notoriety came not only from its unprecedented doors and exceptional performance but for its many innovations. These inspired *Sports Car Illustrated* to proclaim, "Literally, the 300 SL is a car of the future that can be possessed today."

THE GROUNDBREAKERS
FERRARI 250 GTO

FUN FACT

The signature styling feature of the Ferrari 250 GTO is its lengthy nose, which is tipped with distinctive air intakes. Due to production variances, the configuration of these intakes varied from car to car.

HISTORICAL TIDBIT

Ferraris are archetypically red, but the most notorious 250 GTO was, in fact, green. Stirling Moss's verdant GTO was built for him prior to his career-ending crash in 1962; it sold for $35 million in 2012.

KEY PERSON

Enzo Ferrari was a famously mercurial leader, shifting his alliances at the drop of a hat. A notorious mutiny of key collaborators in 1962 (later known as "The Purge," "The Palace Revolt," or the "Ferrari Walkout") turned the tables on Enzo with the departure of 250 GTO impresarios Giotto Bizzarini (who defected to Lamborghini) and Sergio Scaglietti.

If mystique and rarity are the measures of a supercar, the Ferrari GTO might just claim ultimate status among its peers. Short for Gran Turismo Omologato, the most famous GTO of all was originally intended as a homologation special to satisfy the Federation Internationale de L'automobile (FIA's) requirement of building 100 street cars to qualify the model for competition. Enzo Ferrari only built 39 GTOs, allegedly skipping serial numbers and shifting locations of cars in order to elude FIA inspectors.

The 250 GTO, at least initially, claimed a dream team of Italian masters behind the scenes: engineering development was led by the brilliant Giotto Bizzarrini, and the curvaceous bodywork was designed by the legendary Sergio Scaglietti and his team, who hand-hammered the aluminum body into shape using wooden forms called bucks.

Virtually none of the first thirty-six cars produced between 1962 and 1964 were identical due to the vagaries of hand craftsmanship and constant modifications made to the vehicle's setup. But they all featured a tubular steel frame and an uprated V-12 that was the stuff of legend: the 3.0-liter engine produced a bellowing 300 horsepower at a screaming 8,400 rpm (the 250 model nomenclature referred to the 250cc displaced by each individual cylinder, though three four-liter cars were built and referred to as 330 GTOs). Linked to a new five-speed synchromesh gearbox, the GTO was remarkably reliable for its time, which might have had something to do with its participation in more than five hundred races, where it scored victories in everything from outright victories in the Tour de France to class wins at the Targa Florio and Le Mans.

Rarity and desirability have also played a key role in the 250 GTO's stratospheric resale values: the GTO has become the holy grail for Ferrari collectors, breaking its own record in 2013 as history's most expensive car with a sale price of $52 million.

THE GROUNDBREAKERS
FORD GT40

FUN FACT

Though its official moniker was GT, Ford's race car earned the "GT40" nickname because its roofline sat a mere 40.5 inches above ground. The street cars actually measured between 45 and 46 inches tall.

HISTORICAL TIDBIT

Just over 100 road-going examples of the 1960s-era GTs were built, but Ford returned to the street in 2005. The carmaker revisited the GT yet again in 2017 with a homologation model whose race counterpart won its class at Le Mans in 2016 on the fiftieth anniversary of the GT's original 1-2-3 Le Mans finish.

KEY PERSON

After hiring Lola to build the first GT40s in 1964, Ford turned the project over to legendary tuner Carroll Shelby the following year, though it wasn't until 1966 that the father of the Cobra scored Ford its long-awaited win at Le Mans.

Among the most epic automotive rivalries is the Ford/Ferrari feud of the 1960s, which spawned the creation of the Ford GT40. The contention began when, after expressing interest in selling his company to the Ford Motor Company, Enzo Ferrari balked upon learning that his cars would be banned from the Indianapolis 500 race to avoid competing against Ford's Indy cars.

Henry Ford II, outraged by Ferrari's snub, responded by saying, "All right, if that's the way he wants it, we'll go out and whip his ass." Ford plotted his revenge at the highest profile motorsports stage on the planet, the 24 Hours of Le Mans, in Sarthe, France.

Because Ford had no experience in the highly specialized realm of endurance race cars, the manufacturer hired UK-based firm Lola to build the race car, dubbed the GT (for Grand Touring). While Ford delivered a strong debut effort at 1964's Le Mans race, none of the three cars entered finished the race, with victory proving elusive again in '65. History wasn't made until the following year, when Ford handed team management duties over to Carroll Shelby, who helped secure a stunning 1-2-3 photo finish, the first American win at a major European race in four decades. The knife was further twisted on Ferrari when GT40s took Le Mans victories in 1967, 1968, and 1969.

The low-slung, road-going GT40 was built to meet racing homologation requirements, and its race-developed underpinnings made it a wildly effusive alternative to its street-legal contemporaries. Kitted with a plethora of scoops and vents to manage airflow, the street-legal GT40—of which only 107 were made—was essentially a detuned, road-ready version of the race car. The GT40 was initially powered by a mid-mounted 289-cubic-inch V-8. Ford built 31 examples of the Mark I GT40 road car and a grand total of just over 100 street models, with later 427-cubic-inch models more closely resembling the rumbling hellraisers that won at Le Mans.

THE GROUNDBREAKERS
LAMBORGHINI MIURA

The dictionary entry for the word *supercar* might as well be accompanied by an illustration of the outlandish Lamborghini Miura, which is widely considered the first car to legitimately deserve the title. Debuting as a rolling chassis at the 1965 Turin Auto Show, the Miura claimed several unprecedented packaging innovations, most notably a midengine, rear-wheel-drive configuration featuring a transverse- (sideways-) mounted V-12.

Built a mere three years after industrialist Ferruccio Lamborghini famously founded his company to spite Enzo Ferrari, the Miura was the follow-up to the elegantly understated 400 GT, boasting eccentric body curvatures and expressive eyelash-like headlamp accents. The extreme styling was a shock-and-awe salvo lobbed at Enzo's expanding exotic car empire, as was the Miura's 4.0-liter, 350-horsepower powerplant shoehorned into a car that tipped the scales at less than 3,000 pounds. Thanks to its favorable power-to-weight ratio, the slinky two-seater catapulted to a top speed of over 170 miles per hour. Though its handling was notoriously dodgy, especially at higher speeds where the nose was prone to lift as the fuel tank emptied, the Miura's striking appearance and charismatic personality went a long way toward endearing it to royalty, rock stars, and bon vivants.

The Miura's complex engineering was masterminded by Gian Paolo Dallara, an alum of Ferrari and Maserati who would later go on to found Dallara Motorsports and is a successful race car builder still active to this day. But it was its twenty-five-year-old designer Marcello Gandini, working under the design house Gruppo Bertone, who endowed the Miura with its notorious styling. Widely considered not only the sexiest production car money could buy in its day, the Miura was also the fastest, a fact that helped earn Lamborghini the much-needed notoriety that eventually put the company on the map. Though Ferrari countered with the V-12-powered Daytona, a proper retaliation didn't come until the 1973 debut of the flat-twelve-cylinder-powered 512 Berlinetta Boxer, Ferrari's first mid engine production car.

THE GROUNDBREAKERS
FERRARI 365 GTB/4 DAYTONA

FUN FACT

The Daytona has made numerous cameos in pop culture and cinema, but much of its lore stemmed from an appearance in 1971's inaugural Cannonball Baker Sea-to-Shining-Sea Memorial Trophy Dash, also known as the Cannonball Run. Co-driven by event founder Brock Yates and racer Dan Gurney, the Daytona finished the cross-country contest in 35 hours, 54 minutes, achieving speeds of up to 172 mph.

HISTORICAL TIDBIT

While the Daytona's Pespex-covered headlights offered a sleek packaging solution, they didn't meet US federal height requirements. Consequently, retractable headlamps were specially developed for the North American market.

KEY PERSON

Much of the Daytona's aesthetic and functional success can be credited to Sergio Pininfarina's painstaking work at the Turin Polytechnic Institute's wind tunnel.

The supercar arms race was unofficially ignited in 1966 by the Lamborghini Miura, and many considered the 1968 Ferrari 365 GTB/4 to be the answer to Lamborghini's evocative two-seater despite the Ferrari's stubborn adherence to a front-engine configuration. The 365 GTB/4 "Daytona," nicknamed after the American race circuit where the carmaker enjoyed a 1-2-3 finish in 1967, took a different styling approach than its shorter and rounder predecessor, the 275 GTB/4. Featuring a larger 4.4-liter V-12 (the "365" refers to each cylinder's displacement in cubic centimeters), the Daytona expanded its performance envelope with quicker acceleration and a (barely) Miura-beating top speed of 174 miles per hour.

While Enzo Ferrari distrusted aerodynamics, suggesting that horsepower could make up for any drag-creating airflow issues, Daytona designer Sergio Pininfarina made several concessions to the wind tunnel, among them a smaller front air intake, headlights hidden behind a Perspex (transparent plastic) shield, and a swept-back, flush-fitting, double-curved windshield.

The $20,000 Daytona was the most expensive new Ferrari to date, putting it on par with the Miura. While the Daytona's front-engine layout lacked the panache of Lamborghini's radical mid-engine V-12 setup, the long-nosed two-seater was still the fastest Ferrari to date. Its exemplary performance might explain why the faithful Tifosi snatched up 1,383 hard-top models, far eclipsing the 275 model's 200 total units. The success of the Daytona coupe led to an open-air Spyder spinoff in 1969, which irked some purists since its aerodynamic drag erased much of the work that made the coupes so slippery. Only 122 Daytona Spyders were built, 96 of which were sold in the United States. Interestingly, the rarity of the factory-issued Spyder models inspired many coupe owners to convert their cars into droptops.

Regardless of configuration, many die-hards consider the Daytona to be the last great Ferrari, though the 512 Berlinetta Boxer, Ferrari's first mid-engine production sports car, renewed the debate when it debuted in 1973.

THE GROUNDBREAKERS
LAMBORGHINI COUNTACH LP400

While the Lamborghini Miura wowed the world with its feminine lines, eyelash headlamps, and lavish curves, the Countach LP400 followed up with a decidedly abstract and modernist visual treatment consisting of copious trapezoids, ducts, and vents. Hiding the Miura's 375-horsepower, 4.0-liter V-12 powertrain beneath its dramatically wedge-shaped silhouette, the LP400 was an entirely different beast from designer Marcello Gandini, one which eschewed its predecessor's overt sexiness for something more alien and extreme.

The look was so exaggerated that it inspired Nuccio Bertone, upon first seeing the prototype, to exclaim, "Countach!" The expression, uttered in the local Piedmontese dialect, was slang for a sense of astonishment, usually directed toward the opposite sex.

The "LP" in LP400 stood for *longitudinale posterior*, signaling that the massive powerplant was positioned lengthwise, not sideways as in the Miura. This arrangement helped realize Ferruccio's goal of reducing cabin noise, reflecting hints at a desire to make his spacecraft-like creations ever more usable in the real world. But fear not: the waist-high LP400 still had no shortage of otherworldly character to distinguish it from the crowd. Decked out with the brand's first use of scissor-style doors, the early Countach was Lamborghini's most overt expression of defiance to date of the supercar status quo.

The LP400 also claimed more than its share of trick engineering, including a strong and stiff tubular chassis, alloy bodywork, and a gearbox that resided snugly between the two passengers. Only 157 LP400s were built by the factory, making this inaugural Countach the most desirable. Lamborghini followed up with the bulkier LP400 S, which featured massive fiberglass fender arches hiding fat Pirelli P7 tires. Though the most identifiable Countaches are the later big-fendered models, the LP400's clean minimalism most strongly captures the imagination of the purists.

THE GROUNDBREAKERS
PORSCHE 930 TURBO CARRERA

Though the oil crisis and emissions crunch made the 1970s a mostly abysmal decade for supercars, the so-called malaise era was not without its bright spots. Witness, for instance, the Porsche 930 Turbo Carrera, an exotic whose punchy powerplant bucked the downsizing trend while packing enough unpredictable handling quirks to keep things more than a little interesting.

Using Porsche's 911 platform and based on the RSR race car, the turbocharged 930 model was introduced in the US market in 1976 with a then-shocking sticker price of $26,000. While the Porsche nameplate carried with it no shortage of racing cred (especially after the fearsome 917K scored the brand's first victory at Le Mans in 1970), Porsche's notoriously level-headed approach couldn't compete with the raucous V-12s and outrageous styling of its Italian rivals. Instead, the brand focused on daring engineering and efficient packaging that made it a force to be reckoned with on the racetrack.

The wide-hipped, whale tail–spoiler-equipped 930 Turbo Carrera was unlike any previous road-going Porsche, because it finally abandoned the brand's left-brain thinking and dipped its toe into the pool of wretched excess. During an age when other carmakers resigned their performance models to smaller, naturally aspirated powerplants, the 930 went big, with a turbocharger forcing air into the engine to heighten acceleration. Not only was the turbo lag—that is, the pause between the time the throttle was pressed and turbo boost was produced—difficult to anticipate, early Turbo models suffered from trailing throttle oversteer, a trait that triggered a severe oversteer (and often, spinning) phenomenon if the driver lifted the gas pedal mid-corner. The Turbo's tendency to swap ends earned it a reputation as a widowmaker and led many a spirited driver into guardrails and other obstacles.

While the uninitiated may have deemed the Turbo Carrera too similar to its more commonplace 911 stablemates, enthusiasts knew it as the wild-eyed Porsche that could beat many an Italian exotic at the dragstrip and whose ragged-edge dynamics demanded respect.

FERRUCCIO LAMBORGHINI

The most notorious Taurus in the supercar industry was Ferruccio Lamborghini, an industrious enthusiast for all things fleet who built his empire around the leitmotif of bulls.

Lamborghini's agricultural origins exposed him to all manner of bovine specimens as a child, but his interests diverged towards the inner workings of farm equipment. Lamborghini's mechanical aptitude eventually led him to capitalize on Italy's postwar boom by opening a garage where he worked on cars, trucks, and tractors. It was there where he modified a Fiat 500, which he eventually campaigned in the 1948 Mille Miglia and crashed into a wall, spooking him away from racing for life.

Lamborghini's fortunes grew when he started a tractor business, whose success enabled him to outfit his garage with everything from a Mercedes-Benz 300SL Gullwing to a Jaguar E-Type. His purchase of a Ferrari 250GT, 250 SWB Berlinetta, and a 250GT 2+2 enraptured him with the prancing horse brand, which was based in nearby Modena. However, the brand's frequent clutch failures and Enzo Ferrari's dismissal of his complaints led Lamborghini to build a better mousetrap himself, first modifying the Ferraris and then entering the car business in 1963 by forming Automobili Lamborghini in Sant'Agata, Italy.

"It's very simple," Lamborghini said. "In the past, I have bought some of the most famous Gran Turismo cars, and in each of these magnificent machines, I have found some faults. Too hot. Or uncomfortable. Or not sufficiently fast. Or not perfectly finished. Now I want to make a GT car without faults. Not a technical bomb. Very normal. Very conventional. But a perfect car. "

Lamborghini's endeavors initiated with the elegant 350 GT and became wilder with each model (which was subsequently named after a bull), among them unforgettable extroverts such as the Miura, Countach, and LM002.

Lamborghini ran his tractor business alongside his car-making venture, which exposed him to the vagaries and shifting fortunes of two tumultuous industries. Facing financial difficulty in 1974 on the heels of the oil crisis, Ferruccio sold his remaining stake in Automobili Lamborghini, shifting his attention to other business ventures before retiring to a 750-acre farm in Umbria to pursue hunting and winemaking. Though his company had more than its share of ups and downs, including brief ownership by the American Chrysler Company, Lamborghini would have no doubt been pleased with the more recent success of his brand, which is currently owned by the Volkswagen Group's subsidiary, Audi.

GLOSSARY

DIRECT INJECTION: A more sophisticated form of fuel injection that involves spraying a high-pressure stream of gasoline directly into the combustion chamber, allowing more precise control of the combustion process.

FIA: Short for Fédération Internationale de l'Automobile, which serves as the governing body for all international automotive motorsports.

HOMOLOGATION: A set of guidelines in the racing world intended to regulate the rules of participation, as well as a minimum number of sales of equivalent road cars to the public. Homologation often helps trickle racing technology into road-going cars, while keeping the race cars relatable to fans.

MILLE MIGLIA: Italian for 1,000 miles, referring to an endurance race through Italy that was held on public roads from 1927 to 1957. The race saw feverish speeds by enthusiastic drivers, the most notorious of whom was Stirling Moss. Along with his navigator Denis "Jenks" Jenkinson, Moss set an average speed of 97.96 miles per hour over a 992-mile distance. The event is now held as a time-speed-distance event limited to pre-1957 cars.

NATURALLY ASPIRATED: A type of engine that does not use artificial means (i.e., turbocharging or supercharging) to force air into the engine.

OVERSTEER: A condition of vehicle dynamics in which the rear wheels slide, typically due to throttle-induced wheelspin, causing the vehicle to turn more than intended.

SYNCHROMESH: A mechanism within manual transmission gearboxes that keeps gears from grinding by matching the cog speeds before they engage, or mesh.

TIFOSI: A term that describes Italian fans, which are often passionate in their support of a certain manufacturer—typically, Ferrari.

TURBOCHARGING: A means of extracting more power from an engine by using an exhaust-driven turbine to pump more air into the combustion chamber.

TUBE CHASSIS: A type of chassis that uses a complex network of tubes, usually steel or aluminum, to form a car's body. Early supercars embraced this technique despite its labor-intensive assembly process because it enables construction that is both light-weight and strong.

WIND TUNNEL: A structure used to measure aerodynamic efficiency. This methodology became crucial in the supercar world because it helped refine key attributes, such as drag and high-speed stability.

EYE FOR DESIGN

EYE FOR DESIGN
DANGEROUS CURVES

The early days of supercars were a free-spirited era when form ruled supreme over function. Empowered to dream up exuberantly expressive vehicles, stylists penned whimsical rolling sculptures that also happened to fly down the road. As a result, practical concerns such as comfort, handling, and high-speed stability often took a back seat to beauty.

Those formative years were heavily influenced by the 1960s-era Lamborghini Miura, the car many credit for pioneering the supercar craze. Nary an inch of its decidedly feminine body could be faulted; from its just-so stance to its soothing proportions, the Miura exuded a physical presence informed by high design. However, those arbitrary aerodynamics also tended to lift the front end at high speeds, making it a handful for even experienced drivers. Though the Miura was capable of over 170 miles per hour, it would take a foolhardy soul to attempt such velocity.

Not all was lost to impracticality, though. The Miura's endearing "eyelashes" around its headlights doubled as air ducts, directing cooling air to the front brakes. For the most part, however, this exotic's delicate curves and details suggested that anything could happen (be it skittish handling or spontaneous combustion), adding to both the car's inherent allure and its intrinsic danger.

Subsequent cars, such as the Ferrari 250 GTO and Shelby Cobra 427, came to define the bronze age of the supercar, though science and engineering—not to mention safety and government legislation—eventually crept into the equation, forming a dynamic tension between adrenalized absurdity and basic functionality. Advances in aerodynamics using wind tunnels (and later, fluid dynamics, or computerized simulations) put more control in the driver's hands, enabling cars to travel faster with greater stability. Thanks to the eventual introduction of safety features such as crumple zones, fuel cells, and airbags, the inevitable exotic car wrecks would bruise egos more than they did bodies.

EYE FOR DESIGN
THE WEDGE

FUN FACT

Wedge designs typified 1970s supercars, but the movement eventually subsided as other designs gained popularity. One of the few manufacturers clinging to the wedge shape is Lamborghini, whose Aventador and Huracan models still adhere to the doorstop theme.

HISTORICAL TIDBIT

While style-savvy Italians all but invented the wedge, the Germans offered their own functional twist. The Mercedes-Benz C-111 was introduced at the 1969 Frankfurt Motor Show and featured a fiberglass body, gullwing doors, and a rotary engine. The C-111 set numerous sustained top-speed and endurance records that went unbroken for decades.

KEY PERSON

The undisputed godfather of the wedge is Marcello Gandini, who penned the Lamborghini Marzal, Alfa Romeo Carabo, Lamborghini Countach, and Lancia Stratos.

Of all the body shapes embraced by supercars, few exemplify the genre so crisply as the wedge. The mythology of the fastest, wildest, and most memorable cars is steeped in the wedge's inherent attractiveness: Pointy at the nose and escalating in volume toward the rear, the wedge embodies a clean, angular form, slicing through air like a blade.

Wedges evolved out of curvaceous 1960s-era supercars, whose forms seemed to reflect the experimental, boundary-busting spirit of the era. The subsequent linearity was a shock of the new—the prevailing organic shapes of the automotive world became stale, and the introduction of the sharp, "folded paper" aesthetic defied the preceding curves with striking modernity.

The first noteworthy wedges emerged by the late 1960s in the form of competing Italian concept cars from the Bertone and Pininfarina design houses. Bertone unveiled the Lamborghini Marzal in 1967, a Marcello Gandini–designed concept whose shape would later inspire the Espada. The Alfa Romeo Carabo debuted at the 1968 Paris Motor Show and was as wedge-y as a space-age doorstop. Also designed by Gandini for Bertone, the Carabo sat 39 inches high and featured scissor doors (which would later appear on the Countach). The 1969 Turin Motor Show saw Pininfarina's debut of the Ferrari 512S Berlinetta, penned by Filippo Sapino, who would go on to design the inspired 365 GTC/4. The following year, Paolo Martin's wildly flush and linear Ferrari 512S Modulo broke new ground for Pininfarina with its futuristic silhouette, seemingly drawn with one line. Bertone's salvo was the Lancia Stratos HF Zero, an even more outlandish take on the wedge that essentially consisted of a creased snout, a lifting hatch that revealed a cavity that doubled as a cockpit, and a chevroned engine cover that could have served as a *Star Trek* set piece.

The rally-homologated 1973 Lancia Stratos HF grew directly from the Zero's design, while the 1974 Lamborghini Countach LP400 elaborated on the theme established by the 1971 LP500 concept car. Other well-known wedges include the Lotus Esprit, De Tomaso Pantera, and Aston Martin Lagonda.

EYE FOR DESIGN
SCOOP, DUCTS AND VENTS

To help achieve their seemingly supernatural performance targets, supercars are equipped with a symphony of mechanical systems that enable them to out-accelerate and out-corner more pedestrian counterparts. And while scoops, ducts, and vents look undeniably cool—enough to be mimicked in lesser cars that don't necessarily need them—these small features play a crucial role in helping supercars achieve their performance potential.

Aerodynamic elements such as tapering appeared as early as the 1920s, but specific lessons from the wind tunnel didn't make their way into production cars until later. It wasn't until the 1950s that aero technology inherited from race cars trickled down to street cars, and only in the late 1960s did some of the more stubborn manufacturers begin shaping their sheet metal to follow aerodynamics principles.

When it comes to scoops, ducts, and vents, the name of the game is airflow—specifically, the management of air so it achieves the sometimes-contradictory functions of allowing air to reach the engine, gearbox, and brakes, while also allowing air to exit where necessary, all while slipping through the atmosphere as efficiently as possible. Because the powerful energy-dense engines powering supercars are prone to produce heat, cooling becomes another crucial consideration. Internal cooling is handled by liquid coolant, among other methods, while external cooling can be facilitated by outside airflow. Because drawing air into the body can disrupt airflow and cause drag, some carmakers use NACA ducts instead of simple inlets, because they interfere less with aerodynamics. However, NACA ducts suffer the side effect of reducing airflow rates when compared to more standard inlets.

In addition to the intake of air into the body for cooling everything from engine radiators and oil coolers to brakes and transmissions, carmakers also engineer ways to extract air, which evacuates heat and can also dissipate turbulent air created by rotating wheels. Twenty-first-century supercars take these considerations a step further by actively closing ducts and vents for reduced drag or when cooling is not needed.

EYE FOR DESIGN
THE DOWNFORCE GAME

Supercar tuning is in a constant state of tension between downforce and outright speed. Downforce represents one pole of desire because it enhances handling; the more aerodynamic downforce is generated, the greater the grip (and therefore, the better the handling). On the other hand, downforce creates drag, which has the un-supercar-like side effect of slowing the vehicle.

Early supercars were designed with little regard for downforce. As such, these beautiful brutes often had blistering acceleration, impressive top speeds, and the tendency to lose their grip on the road. Looking to the world of racing during the so-called "ground effects" era of the 1960s, the other extreme was also encountered: such absurd downforce that suspension systems strained to the point of failure, causing wishbones and control arms to snap and send cars spinning out of control.

As supercars of the 1960s and beyond inherited spoilers, splitters, skirts, and diffusers from race cars, they not only benefited from the performance gains of these devices, they also assumed their butch, no-nonsense looks. The badassery of this bodywork reflected its function, but it wasn't long before the race-derived hardware found on automotive exotica trickled down to more attainable performance cars and even passenger cars.

As with the balancing act between ducting air into the body for cooling and minimizing turbulence and drag, the tuning of downforce is often a push/pull between opposing results. Creating an optimal setup biased toward top speed can make for a car that feels skittish the faster it goes; likewise, too much downforce can sacrifice acceleration and top speed—prime bragging points for supercars—as well as put undue strain on suspension components, which will then require heavier construction. The development of active spoilers and flaps, which activate based on speed-calculated algorithms, goes a long way toward ensuring that optimal downforce keeps the vehicle in control without excessively hampering velocity.

EYE FOR DESIGN
FORM BATTLES FUNCTION

Extraordinarily fast, visually arresting cars have rarely been engineered for workaday features such as cargo capacity or fuel economy. But a decades-long trend has gradually seen supercars evolve from unreliable single-purpose exercises in impracticality to surprisingly comfortable, capable, and versatile machines.

The era of carburetion and curves—think AC Cobra and Ferrari 275 GTB—witnessed the supercar as an exercise in compromise. Driving an early supercar often meant your steed was obnoxiously loud, unhappy at low speeds, and temperamental. As carburetors evolved into fuel injection and suspension systems became able to automatically soften or stiffen their damping, drivability improved dramatically. Hardware and engineering advances inherited from racing also improved the breed, enhancing durability and reliability. While complex tube chassis went the way of the dinosaur, unibody construction increased cabin volume, enabling stiffer structures that facilitate both smoother rides and more responsive handling.

As the livability of supercars continues to improve, carmakers feel increasing pressure to differentiate their offerings in the face of stiff competition. If a car's performance is comparable to that of its competitors, would-be buyers are likelier to choose their next ride based on differentiators such as design and exclusivity. Also more crucial during the modern era are mechano-emotional elements such as exhaust sounds, engine configurations, and that ever-elusive x factor of driving feedback. With each manufacturer bringing its own unique set of characteristics to the table— be it technology, country of origin, or race heritage—the landscape is in a constant state of flux.

Brands are faced with the challenge of staying true to their heritage while achieving ever-higher performance targets. The technologies that enable an ever more comfortable and safe driving experience also threaten to diminish its excitement—is a supercar *super* if it is comfortable and quiet enough to serve as a grocery getter?

EYE FOR DESIGN
BIOMIMICRY INCARNATE

From honeycomb patterns in composite structures to spoilers that echo the shapes of bird wings, modern supercars borrow countless cues from nature in pursuit of greater efficiency and speed. The practice dates to the 1930s, when teardrop-shaped cars sought to cheat the wind. Today's designers and engineers take a far more targeted approach, examining everything from micro bends and contours of wing shapes to microscopic patterns in fish scales.

While supercar designers of the 1950s and 1960s seemed intent on personifying their creations with anthropomorphic "faces" and curvy bodies that closely resembled the female form, it wasn't until the widespread reliance on wind-tunnel data (and later, computational fluid analysis) that the benefits of biomimicry became explicitly clear. The practice of borrowing from nature isn't unique to automobiles; witness the proliferation of winglets, the small projections on the tips of aircraft wings, which was inspired by the way the wing tips on birds of prey fold up in order to reduce drag. Biomimicry can also be found in disciplines ranging from aircraft design to architecture.

Most of today's work in biomimicry is aimed at increasing aerodynamic efficiency, reducing wind noise, and efficiently diverting airflow. Consider, for instance, the five small bumps on McLaren's mirror stalks: inspired by the fast-moving sailfish, they divert enough airflow to considerably reduce cabin noise. Applying that same creature's scale texture to the engine intake ducts of their P1 enabled 17 percent more air to enter the combustion chamber.

If concept cars are any indication of the future of biomimicry, brace yourself for advanced ideas such as shapeshifting body panels, which anticipate high speeds by tucking into a sleeker shape.

EYE FOR DESIGN
SUPER SUVs

FUN FACT

Critics taking aim at the Urus, Lamborghini's sophomore effort at the SSUV genre, will struggle finding fault in its performance; the 189-mph Urus is capable of catapulting from 0 to 62 mph in 3.6 seconds, which is quicker than the Gallardo, Lamborghini's previous generation entry-level supercar.

HISTORICAL TIDBIT

The Lamborghini LM002's roots as a military concept informed its stripped-down styling, but its "Rambo Lambo" nickname stemmed from LM002 owner Sylvester Stallone, who starred in the Rambo action films.

KEY PERSON

Ferrari has resisted venturing outside the supercar genre, but CEO Sergio Marchionne marked a dramatic break from tradition by finally admitting the brand is working on an SSUV. "By definition it's going to drive like a Ferrari, it has to," Marchionne says of the impending Super SUV.

Supercars are usually exactly that—cars. But the emergence of the so-called Super SUV (SSUV)—high-riding, plus-sized vehicles with exceptional performance and luxury—flips that convention on its head.

The SSUV came on the heels of the meteoric rise of sport utility vehicles in the early 2000s. As premium SUV sales eclipsed those of sedans, brands such as Range Rover capitalized on the trend with bespoke offerings whose interiors rivaled those of tony British brands, such as Rolls-Royce.

The commercial viability of this growing niche was paved by the Porsche Cayenne, whose 2003 introduction initially infuriated brand fanboys. The Cayenne proved its mettle by eclipsing the sales of Porsche's venerable 911, and quite possibly saving that flagship model from extinction. In 2015, Bentley introduced the Bentayga SUV, a twelve-cylinder, 187-mph ultraluxury vehicle whose blend of performance and plushness surprised critics; Lamborghini followed suit with the debut of the Urus, their own 641-horsepower incarnation of a supercar on stilts.

Though SSUVs have entered the enthusiast consciousness relatively recently, ground zero for the genre can be traced back to 1977, when Lamborghini built the Cheetah, a rear-engine military-vehicle prototype that failed to win a government contract and nearly bankrupted the company. Another offroad Lamborghini concept dubbed the LM001 debuted in 1981, but it wasn't until 1986 that Lamborghini's first production SSUV appeared at the Brussels Motor Show, the LM002. The shockingly angular vehicle embodied excess. With a hulking Countach V-12 engine housed under its domed hood, the three-ton monster combined menacing, post-apocalyptic styling with brawny performance and a leather-swathed interior. Only 300 LM002s were built between 1986 and 1992, putting a nail in the SSUV coffin until the genre's 21st century renaissance.

Posh brands such as Aston Martin and Rolls-Royce now have Super SUVs in the works, as does Ferrari, putting pressure on early adopters with a prancing horse interpretation of the SSUV paradigm.

GORDON MURRAY

Gordon Murray is the father of the McLaren F1, widely considered one of the greatest supercars of all time. Murray's expansive career began as a chief designer for Brabham's Formula 1 team in 1969, a formative time for the industry that enabled him to push the boundaries of motorsport engineering. By designing cars such as the BT55, which managed to achieve high downforce with minimal drag, Murray put himself on the bleeding edge of the sport's considerable talent pool. His experience working for McLaren's F1 team between 1987 and 1991 fueled racing phenom Ayrton Senna's meteoric rise to fame, not to mention teammate and rival Alain Prost's success.

After contributing to four back-to-back Constructor and Driver championships for McLaren, Murray leapt to lead design for McLaren's upstart road car business, McLaren Cars. His first project, the McLaren F1, was a no-holds-barred attempt to build the ultimate road-going supercar. The first production car with a carbon-fiber monocoque construction, the F1 weighed 2,244 pounds—about as much as a wispy Mazda Miata—but was powered by a monstrous 6.1-liter V-12 powerplant that produced 618 horsepower.

No expense was spared in making the F1 exceptional. Gold foil was chosen to insulate the engine, Kevlar fans helped generate downforce, and a plethora of advanced hardware enabled staggering performance numbers: 0 to 60 miles per hour in 3.2 seconds and a record-busting top speed of 240 miles per hour. The F1 only saw 107 total cars produced (71 as road cars), and though its asking price when new was a then-scandalous $815,000, the most recent publicly sold car traded hands for $15,620,000.

Following the F1, Murray worked on the Mercedes-Benz SLR McLaren, a supercar in its own right but one whose collaboration between Mercedes-Benz and McLaren resulted in some inevitable compromises. The long view revealed that the SLR would never claim the legendary status of the F1.

Murray founded his own independent consultant business in 2007, later taking on such diverse projects as an ultra-efficient city car and an ultra-lightweight two-seat performance car.

GLOSSARY

AERODYNAMICS: The study of airflow and its interaction with objects. In the case of vehicles, aerodynamics reflects how much energy it takes a car to cut through air and the relative stability of that car at speed based on the balance of downforce.

BERNOULLI'S PRINCIPLE: Identified by Swiss mathematician Daniel Bernoulli, this theorem outlines how faster moving air results in lower pressure. The phenomenon explains why a wing produces lift and how a vehicle's bodywork can create areas of downforce.

BIOMIMICRY: The incorporation of features found in nature or animals into the design or structure of inanimate objects.

COEFFICIENT OF DRAG: A measure of how much resistance an object encounters while moving through space due to its size, shape, and airflow characteristics.

COMPUTATIONAL FLUID DYNAMICS: A branch of computer simulation that replicates a wind tunnel by treating the movement of air molecules as fluids. Computer simulations have come to replace many of the hours previously spent in physical wind tunnels.

DIFFUSER: A ribbed tray or structure, usually placed at the tail of a vehicle, designed to accelerate airflow beneath a car in order to form a zone of negative pressure that creates downforce.

FLAP: A small, paddle-like wing that can be fixed or actively moved in order to manipulate airflow. The **GURNEY FLAP,** developed by American racer Dan Gurney, is an airfoil added to the trailing end of a wing that aids its performance by enhancing airflow and suction to the wing.

DOWNFORCE: The downward force exerted on a vehicle as it moves. Downforce can help a vehicle travel faster through corners, but the turbulence created by downforce can reduce acceleration and top speed.

KAMMBACK: A body style, named after Wunibald Kamm, in which the tail ends with an abrupt edge. The aerodynamic profile created by the shape reduces drag.

SKIRT: A boundary on the edges of a vehicle that contains air as it passes below to maintain a low-pressure boundary, helping suck the car down and create downforce.

SPLITTER: A horizontal extension at the nose of the vehicle intended to "split" the air, sending high pressure above in order to produce downforce.

SPOILER: A flap that "spoils" or disrupts airflow to prevent aerodynamic lift. Spoilers are not to be confused with wings, which exist to create downforce.

UNIBODY: A chassis structure that uses a unified or one-piece body/frame. Unibody vehicles tend to be structurally stiff and less complex than tubular chassis.

VENTS: Openings in a vehicle's body that enable the evacuation of air, be it for airflow or heat dissipation.

WING: An airfoil-shaped structure that creates downforce by deflecting airflow upwards. Wings on cars act the opposite way they do on aircraft because they create downforce by being mounted upside down.

WIND TUNNEL: A structure that measures aerodynamic drag and lift by sending airflow past a static vehicle.

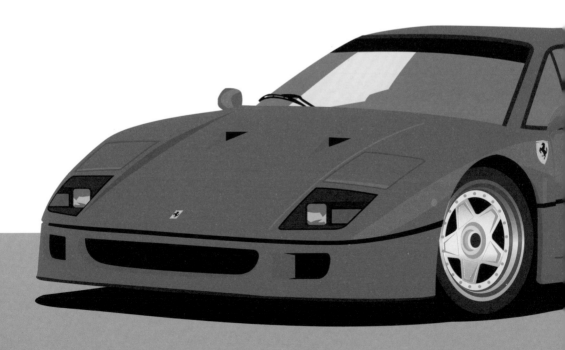

1980s RENAISSANCE

1980s RENAISSANCE
LAMBORGHINI COUNTACH LP500S

FUN FACT

Lamborghini explored increased horsepower by commissioning two LP500Ss with turbocharged power-plants. The experiment never took off, and only one of the two concept cars remains.

HISTORICAL TIDBIT

One of the Countach's most identifiable design features was its massive spoiler, which further amplified the car's already brash presence. Originally commissioned by F1 team owner Walter Wolf, the so-called Wolf Spoiler actually slowed the car down.

KEY PERSON

CEO Ferruccio Lamborghini received most of the glory for creations such as the Countach, but the car couldn't have come alive without the engineering of Paolo Stanzani and Massimo Perenti, and company pioneers such as Gion Paolo Dallara, Giotto Bizzarrini, and Bob Wallace.

In 1974, the original Lamborghini Countach LP400 wowed the world with its insectoid profile and improbable geometry, but the scissor-doored beauty didn't hit its peak excess until the 1980s, a decade that required the Italian manufacturer to step up its game to play hardball with the burgeoning competition. While the earlier LP400S model added hulking fiberglass fender flares and fatter tires (which gave the Countach its signature muscular stance and improved handling), the 365-horsepower variant was actually slower than the preceding model—not very super, for those keeping score at home.

It wasn't until the LP500S came along in 1982 that the bad-boy super-car was endowed with the gruntier 4,754cc V-12, which once again lit a fire with archenemy Ferrari. While the flared fenders boosted its street cred, the LP500S's uprated performance was key to cementing Lamborghini's reputation in the supercar microcosm. Despite the new model's heavier weight, the V-12's 375 horsepower and 302 lb-ft of torque paved the way for the ultimate Countach, the LP5000 QV (or Quattrovalvole), a 455-horsepower screamer whose brutal performance finally matched its otherworldly looks. Also, commensurate with the overstated styling are its driving dynamics: a lack of power-assist steering and a notoriously heavy clutch made the Countach a handful to maneuver at slow speeds, and pleasurably painful to drive fast. Its low-slung layout proved unfriendly to real-world conditions, with limited rear visibility forcing drivers to resort to the tried-and-true method of swinging up the scissor door, sitting on the doorsill, and looking over your shoulder while backing into a parking spot.

Though purists invariably prefer the cleaner lines of earlier models, the fat-fendered, big-winged later variants are what became the stuff of childhood posters and celluloid dreams. It is that oversized persona and larger-than-life presence that makes the Countach among the most quintessential of supercars.

1980s RENAISSANCE
PORSCHE 959

While most 1980s-era supercars gained notoriety through wild styling and crackpot charisma, the Porsche 959 earned its infamy with rigorous engineering and Teutonic values. Based loosely on the brand's decades-old 911 platform, the initial iteration of the 959 was first seen at 1983's Frankfurt Motor Show and referred to as "Gruppe B," the rally car competition class dominated by no-holds-barred monsters, such as the Audi Quattro S1 and the Ferrari 288 GTO.

Group B racing required the production of at least 200 road-going examples. Though it bore a passing resemblance to the 911's silhouette, the 959's Kevlar and aluminum bodywork wrapped around dramatically re-engineered hardware. The 911's air-cooled flat-six was replaced with a high-output 450-horsepower engine that featured liquid-cooled heads and the first sequential turbocharger system. The powerplant mated to a six-speed manual gearbox, driving power through all four wheels via a sophisticated new electronic all-wheel-drive system, which used torque vectoring for sharp cornering. Also aiding handling was a complex active suspension system with user-adjustable ride height and stiffness, which made the 959 versatile enough to be driven on or off paved surfaces. Among the 959's many innovative details were hollow-spoke magnesium wheels, which held air along with the tires, and the world's first usage of run-flat tires.

The first trial for these underpinnings came with the so-called 953, which was used as a development test bed for the 959. Tackling the Dakar Rally, a grueling 7,500-mile race through Africa, the 953 performed flawlessly, finishing in first place—a staggering achievement. The platform's versatility was later proved in 1986 when the 961, a road race–based 959 variant, competed at the 24 Hours of Le Mans and finished first in its class and seventh overall.

With only 337 examples produced, history views the 959 as a rare and exceptional vehicle whose performance benchmarks inspired Ferrari to retaliate with the F40.

HISTORICAL TIDBIT

The Ferrari F40 was the first car to officially break the 200-mile-per-hour barrier (with an official top speed of 201 miles per hour), though magazines had a tough time achieving that figure themselves with independent testing: *Car and Driver* hit 197, while *Road & Track* could only muster 196.

KEY PERSON

The F40 was the last car to receive Enzo Ferrari's official stamp of approval, and il Commendatore died at the age of ninety on August 14, 1988, the same year the F40 hit the market.

You'd never guess the acutely styled Ferrari F40 evolved from the Coke bottle-shaped 288 GTO Group B car, but the beast unveiled on July 21, 1987 was more than just a stunning new model. The F40 commemorated the brand's fortieth anniversary and became the most important road-going Ferrari of its time.

As the last car approved by Enzo Ferrari, the F40 marked the end of a momentous era. Extensive development work went into making it the fastest, most powerful contemporary Ferrari. The F40 went from concept to finished product in a mere twelve months because it was intended as a rapid response to Porsche's 959.

While the 959 was a technical tour de force, the F40 was raw and brutal—a race car for the road. Weighing just over 3,000 pounds, the F40 utilized composite structures bonded to a tubular steel chassis and a total of eleven carbon/Kevlar composite body panels. Key to the F40's design was a severely sloped nose, which reduced its frontal area for less drag. Its track-focused capabilities were punctuated by a massive spoiler.

Creature comforts were spared to fulfill its mission as a relentlessly focused performer: the cabin was devoid of leather, door panels, carpeting, and a glove box. The first fifty cars had sliding Lexan windows, while later iterations used hand cranks. The extreme weight saving was complemented by a twin-turbocharged 2.9-liter V-8 that produced 471 horsepower and enabled an unprecedented 0–60-mile-per-hour time of 3.8 seconds, which beat its tech-laden nemesis, Porsche's 959. Acceleration figures depended largely on driver skill, of course, since gearshifts were through the notoriously tricky gated Ferrari shifter.

The F40, produced between 1987 and 1992, represented the pinnacle of the brand's road cars and a towering monument to Enzo Ferrari's four decades of dominion. With a then-staggering $400,000 entry cost and an initial promise of only 450 models, dealer markups inflated the price to nearly double. Production was later expanded to 1,311 examples to satisfy more of the brand's most loyal fanatics.

1980s RENAISSANCE
FERRARI TESTAROSSA

If the late-80s era Ferrari F40 was the emotional response to the technologically rational Porsche 959, the introduction of the Testarossa in 1984 played a reversed role to its nemesis the Countach, offering a refined retort to Lamborghini's outrageous bull.

The Testarossa ("redhead" in Italian) inherited its name from the curvaceous, 1950s-era racer, the 250 Testa Rossa. A follow-up to Ferrari's first twelve-cylinder midengine car, the Berlinetta Boxer (BB 512), the new model introduced excessive 1980s styling and lavish proportions—particularly its expansive 1,976-millimeter width, which made it the broadest road-going Ferrari for the better part of three decades. Though the Testarossa's signature side strakes appeared decorative, they hid twin radiators that cooled a redheaded 4.9-liter flat-twelve-cylinder powerplant. It was more powerful than the 352-horsepower Berlinetta Boxer it evolved from, but the new car's more efficient packaging also made it a modern exotic that was far more livable, with decent visibility, a relatively well-insulated cabin, and effective air conditioning.

Producing 385 horsepower in its earliest iteration, the Testarossa's engine played yang to the Countach's brutalist yin, a melodious response to the Lamborghini's raucous call. Also harmonious were the Testarossa's driving dynamics, which offered a manageable ride and more predictable handling. A spacious cabin delivered greater comfort and more overall usability, making the Testarossa one of the first functional supercars of its era.

Later iterations included 1992's 512TR, which offered handling refinements and a power boost to 428 horsepower, and 1994's 512M (denoting *modificato*), a 440-horsepower evolution of the breed that served as the swan song for the flat-twelve powerplant. With steep initial demand leading to elevated pricing, the Testarossa proved an outright success during its twelve-year production run, eventually selling a total of 7,177 cars—a rather large figure for a supercar that reveals the Testarossa's all but universal aesthetic appeal.

1980s RENAISSANCE
VECTOR W8

On the heels of the 1980s maelstrom of headline-grabbing supercars hailing from the world's most iconic manufacturers, the Vector W8 emerged as an outsider armed with futuristic styling and powered by American muscle. The brainchild of former Detroit auto executive Jerry Wiegert, the W8 made attention-grabbing performance claims under the guise of aerospace-inspired engineering.

Built in Wilmington, California, the W8 featured a GM-sourced, twin-turbocharged 6.0-liter V-8 that produced a stout 650 horsepower and 650 lb-ft of torque, numbers which far outclassed the day's top machines from Ferrari and Lamborghini. That powerplant was good for a claimed 0–60-mile-per-hour time of 4.2 seconds and an estimated top speed of 218 miles per hour, remarkable figures that eclipsed the achievements of established supercars.

The W8 itself was a construction marvel, featuring an aggressive wedge-shaped body wrapped around a carbon fiber, Kevlar, and fiberglass body. The cabin was upholstered copiously in leather and suede, while the dashboard's small buttons and electroluminescent displays more closely resembled a fighter jet cockpit than a high-dollar sports car. But the Vector's Achilles heel was its modified, GM Turbo Hydramatic-sourced three-speed automatic gearbox. In an age when the most cutting-edge exotics were sporting five and even six-speed manual transmissions, the Vector's three-speed served as a curiously antiquated weak link in an otherwise forward-thinking car.

Only nineteen Vector W8s were produced, with production tapering off when the price escalated from an already considerable $250,000 to a downright ludicrous $400,000. As with most automotive upstarts, Vector eventually went the way of the dinosaur.

1980s RENAISSANCE
BMW M1

BMW's aspirations to join the supercar club were realized with the M1, an elegantly sculpted two-seater. Whether the M1 qualifies for the realm of true supercars or merely high-priced exotics is open for debate, but the M1 was originally conceived in 1975 as a contender for the Group 4/5 World Championship race series, a venue that might have afforded BMW to take a credible stab at Porsche. Unlike most German endeavors, the M1's path to production turned out to be ill-timed and poorly executed.

BMW's plan to build 800 road-going examples far exceeded the minimum homologation requirements of 400 vehicles, a number that turned out to be overly ambitious given BMW's internal production capacity. The subsequent plan was to subcontract the project to an external manufacturer. Initial intentions to outsource manufacturing to Lamborghini fell apart over quality control concerns and the Italian brand's impending insolvency. BMW resorted to a circuitous and needlessly complex manufacturing chain that involved several suppliers across more than one continent. By the time this labyrinthine process was completed, Group 5 homologation requirements had changed, leaving BMW with a vehicle built to a defunct specification.

With a completed car ready to race but no series to run it in, BMW decided to set up its own: Procar, a championship that worked in association with the top tier of automotive racing, Formula One. While legends such as Niki Lauda and Nelson Piquet won titles in 1979 and 1980, the M1 also garnered negative press due to its burgeoning reputation for engine problems. The model competed at Le Mans between 1980 and 1986, though it proved outdated amongst a field of ever more competitive models.

The street-going M1 was expensive for the time at around $55,000, and only 399 road cars and 54 race machines were built. However, future BMW models, such as the E28 M5 and M635CSi, would benefit by inheriting its 273-horsepower inline-six powerplant. As with many things in life, absence has made collector's hearts grow fonder, with a renewed interest in the M1 driving up resale values considerably.

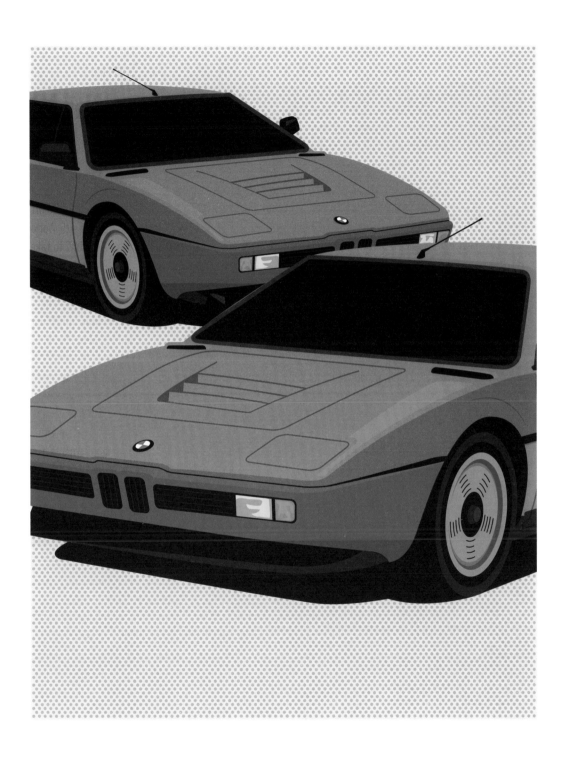

1980s RENAISSANCE
LOTUS ESPRIT TURBO

FUN FACT

Lotus chassis guru Roger Becker stepped in during the filming of *The Spy Who Loved Me* to pilot the Esprit Series 1 when stunt drivers couldn't get the car to perform as needed.

HISTORICAL TIDBIT

The Esprit earned global notoriety when a Series 1 car was featured in 1977's *The Spy Who Loved Me*, the James Bond film in which the car transformed into a submarine. The white Series 1 car from the film was purchased by rocket and electric car impresario Elon Musk in 2013. A 1981 Esprit Turbo was also featured in *For Your Eyes Only*.

KEY PERSON

Lotus founder Colin Chapman was a notorious advocate for "simplifying and adding lightness," a maxim that saw great dividends with the Esprit, which enjoyed a lengthy production run from 1976 to 2004.

Of all the four-wheeled wedges to populate the supercar universe, the Lotus Esprit might have been the most improbable. While the British manufacturer certainly had the chops thanks to its countless innovations within the world of Formula 1 racing, the Esprit began life in 1976 as a small, meagerly powered four-cylinder car—hardly supercar material. It wasn't until the advent of a later, turbocharged iteration of the Series 2 model in 1980 that the Esprit came into its own as a viable contender against bigger, wilder styled, and more powerful competitors.

Key to the Esprit's David-like ability to battle Goliaths was its potent power-to-weight ratio, which took full advantage of its Lilliputian size and enabled responsive handling and blistering acceleration. By using special manifolds and high-pressure Garrett T3 turbochargers, the tiny 2.2-liter four-cylinder delivered lively throttle response and 210 horsepower, enough to sprint as quickly as some Ferraris; later iterations produced enough grunt to outrun the high-dollar, hyper exotic competition from Italy.

Though the small, turbocharged engine packed within the engine bay of the Giugiaro-designed body became a signature of the Esprit, later reaching its zenith with 1995's 302-horsepower Sport 300 model, the Esprit's final Series 4 iteration saw a twin-turbocharged 3.5-liter V-8 that produced 264 horsepower. Added power and increased reliability made the S4 models desirable, but many hold the earlier turbocharged four-cylinder Esprits in the highest esteem, praising their so-called "folded paper" designs and minimalist executions.

CARROLL SHELBY

Though Carroll Hall Shelby earned a lifetime of notoriety for his work in the racing and muscle car worlds, his early endeavors as an oil roughneck and chicken farmer signaled personality traits that would define his groundbreaking accomplishments later in life.

Shelby's automotive endeavors launched with his racing career, which took him from driving diminutive Allards to racing for Aston Martin and Maserati factory teams in the 1950s. Though his successes led him from the Bonneville Salt Flats to Le Mans, a bad heart took him out of competition and forced him into his defining role as the founder of Shelby-American.

He became a supercar legend by outfitting the British AC Ace with American V-8 powerplants to create the legendary Shelby Cobra. Wishing to take on fearsome Ferraris in higher speed endurance races, he and designer Pete Brock conceived the Shelby Daytona Cobra Coupe, of which only five were built. The racing success and rarity of these coupes have pushed their values to astronomic levels.

Among Shelby's more memorable collaborations was his work with the Ford Motor Company to create the Ford GT race car and eventually pull off a stunning 1-2-3 finish at the 24 Hours of Le Mans in 1966, vanquishing archenemy Ferrari in the process.

Shelby would later collaborate with Ford on iconic special edition cars, such as the GT350R and the GT500KR. Though he ventured to build the Series 1, his own ground-up sports car, Shelby's most notorious creations were the ones in which he used his know-how and pluck to concoct memorable performance cars capable of beating the world's most fearsome competition.

GLOSSARY

CARBON FIBER: A strong, lightweight material used in the construction of high-performance automobiles. Up to five times stronger than steel, carbon requires a labor-intensive bonding and curing process to ensure proper structural integrity. Its usage originated in the aerospace industry and was quickly adopted by Formula 1 race constructors. Today, carbon fiber is becoming increasingly popular in mainstream cars.

GROUP B: First introduced in 1982 by the Fédération Internationale de l'Automobile (FIA), Group B racing consisted of a set of regulations that encouraged virtually unrestricted powerplants to be used in rally cars. Frequent deaths and injuries eventually brought an end to Group B's dangerous but sensational run.

FIBERGLASS: A sheet-like material based on a plastic matrix. Weaker but more affordable than carbon fiber, fiberglass was often used to complement more exotic materials.

GM HYDRAMATIC: An extremely popular transmission first introduced in 1939 by General Motors. Not only were Hydramatics the first automatic gearboxes to be mass-produced, they were also remarkably long lived: versions of Hydramatic transmissions persisted into the 1990s.

POWER-TO-WEIGHT RATIO: A measure, usually in horsepower-to-weight, that indicates a vehicle's performance potency based on the relationship between its engine output and its overall mass. The argument for an effective power-to-weight ratio can minimize the seeming importance of engine horsepower, since lightweight bodies require less power to propel them.

THROTTLE RESPONSE: A measure of the engine's responsiveness to throttle input. In supercars, or any performance cars for that matter, a linear relationship between throttle input and the engine's ability to respond by producing power is desirable.

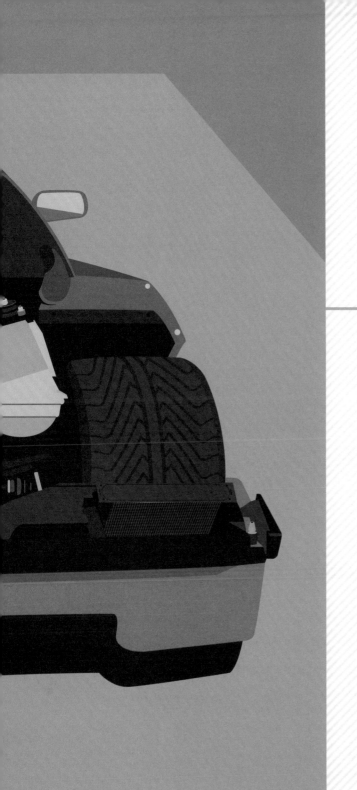

TECH REVOLUTION

TECH REVOLUTION
ENGINE CONFIGURATIONS

No matter how elaborately sculpted its bodywork, the soul of the supercar will always remain its engine. After all, locomotive force is what distinguishes a supercar from lesser vehicles; without that effusive motivation, a supercar becomes just another pretty shape.

That said, there are countless ways to slice a powerplant, the two most immediate being cylinder count and configuration. While some vehicles with fewer than eight cylinders (such as the four-cylinder Lotus Esprit or six-cylinder Jaguar XJ220 and Ford GT) arguably qualify as supercars, the vast majority, at least historically, boast supernumerary cylinder counts. Whether because of the sheer excess associated with a big number or the functional reality that more cylinders often equate to greater power, supercars have been armed with an unusual number of cylinders. That's not to say some legitimate supercars won't claim fewer than twelve, but the mystique behind that number can sometimes mark the tipping point that makes a car a supercar.

Another factor beyond simple cylinder count is their layout, or configuration, which describes how those cylinders are arranged. A straight-eight, for instance, describes eight cylinders laid out in a row, while a V arrangement (V-8, V-12, etc.) lays out the cylinders in a "vee" configuration, with two banks of cylinders usually arranged in a 60-degree or 90-degree setup. So-called flat engines lay cylinders horizontally, usually in banks opposed to each other (the flat-six Porsche 911 or flat-twelve Ferrari Testarossa being examples), while a W arrangement uses narrow cylinder banks with offset double rows to make for more compact proportions (such as the Bugatti Chiron's W16).

Beyond practical considerations, such as size and center of gravity, cylinder counts and configurations also determine an engine's power delivery characteristics, which many enthusiasts recognize as a powerplant's soul. For instance, V-8s are loved for their distinctive rumble and low-end torque, while V-12s are idolized for their silky power delivery and uncanny smoothness.

TECH REVOLUTION
CARBON FIBER CONSTRUCTION

The introduction of carbon fiber construction dramatically altered the supercar landscape, but the shift didn't happen overnight. The technology trickled down from the racing world, where the benefits of the strong, lightweight material revolutionized the sport, enabling better handling, higher speeds, and far greater crash protection.

It took about a decade for the transfer to take place: the first race car with a carbon fiber chassis was the McLaren MP4/1, which raced F1 between 1981 and 1983, and fittingly, the first production car with a carbon fiber chassis came with McLaren's F1 road car. Produced between 1992 and 1998, the F1's carbon fiber monocoque chassis enabled a remarkably light curb weight of 2,500 pounds—about as heavy as a Mazda Miata.

The Mercedes-Benz SLR (2003 to 2010) was a collaboration with McLaren that also utilized a carbon fiber chassis, but its nearly 2-ton curb weight seemed anathema to the intent of the lightweight material. While Ferrari's first road car to embrace the material was the F40 (1987-1992), it wasn't until later that boutique manufacturers, such as Koenigsegg and Pagani explored the technology using hybrid manufacturing techniques that incorporated other lightweight materials, such as strands of titanium woven into the carbon fiber matrix.

Lamborghini came relatively late to the carbon fiber party, which was part of the reason then-engineer Horatio Pagani, an ardent proponent of carbon technology, left to form his own company. While the Lamborghini Countach Evoluzione concept used carbon fiber–reinforced plastic components to save a staggering 1,100 pounds, the material was abandoned until later models could employ a more cost-effective solution.

McLaren's entire lineup of cars now utilize carbon fiber core structures, though Ferrari reserves the material for their most exclusive models—most recent of which is the LaFerrari. Lamborghini is diving deep with carbon, with an in-house manufacturing plant dedicated to composites and a US-based research lab in Seattle, Washington.

TECH REVOLUTION
TURBOCHARGING

FUN FACT

The distinctive whistling and whirring sounds of turbos is typically masked in mainstream production cars but is often unfiltered (and even amplified) in supercars, implying a sense of character and rawness.

HISTORICAL TIDBIT

The twenty-first-century redux of the Acura NSX originally featured a naturally aspirated engine, but when the Japanese carmaker saw supercars embracing forced induction and hybrid power-plants, the NSX's V-6 adopted twin turbochargers.

KEY PERSON

Though turbocharging has swerved in and out of fashion for decades, the technology was invented over a century ago in 1905 by Swiss engineer Alfred Büchi.

Forced induction enhances power output by forcing more air into an engine's combustion chambers. By boosting the airflow entering an engine, usually with a turbocharger or supercharger, more torque and horsepower is extracted, enabling fewer cylinders and/or smaller displacement.

Within the supercar community, forced induction can be a controversial tactic. The practice has roots in aviation, where the technology improves operation at high altitude, because cramming more air into the combustion chamber aids performance when air density is low. Supercharged racers, such as the fearsome Mercedes-Benz Silver Arrows of the 1930s, delivered breathtaking performance, though advances in metallurgy and engine design later enabled free-breathing motors to achieve high-revving power without the aid of forced induction, making naturally aspirated engines commonplace well into the 1960s.

While some of the earliest supercars created their power without turbos or superchargers, those technologies gained credibility when they became commonplace in racing. The early 1980s saw turbocharged Group B rally cars with thrilling (and often lethal) performance and boosted F1 cars surpassing the 1,000-horsepower mark. With those aspirational vehicles achieving huge power through turbocharging, it was inevitable the practice would spread into road cars. One of the most impactful and iconic cars equipped with turbos was the Ferrari F40, whose twin-turbo V-8 produced 471 horsepower—staggering for 1987.

Supercars began employing turbochargers in the 1970s and more heavily in the 1980s, often assuming an added element of excess on top of an already indulgent powertrain—witness the sixteen-cylinder Bugatti Veyron and Chiron, which boast *four* turbos per engine. Massive turbocharged engines power everything from Paganis to Koenigseggs, and in 2014 Ferrari broke a two-decade abstinence from turbos by revisiting the trend across much of its lineup. One holdout? Lamborghini—who, until the release of its twin-turbocharged Urus SUV, has stuck to its naturally aspirated guns.

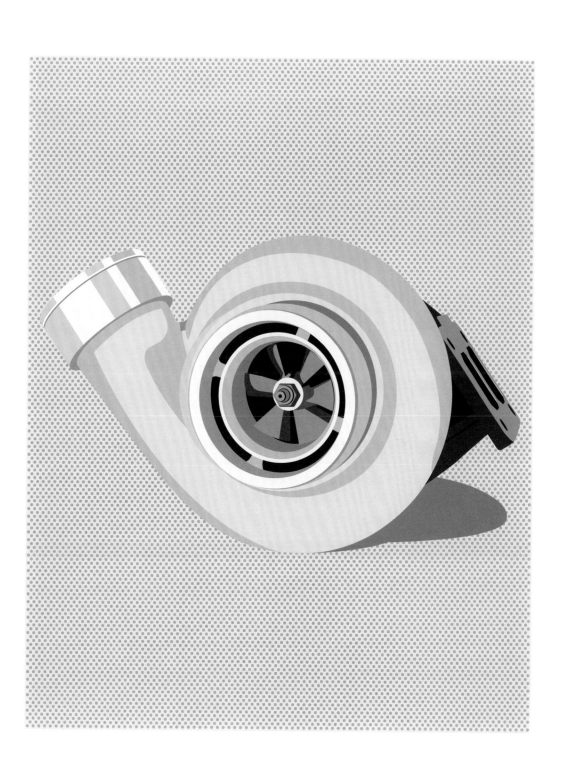

TECH REVOLUTION
CARBON CERAMIC BRAKES

FUN FACT

Carbon ceramic brakes work best at high operating temperatures, upwards of 1,000 degrees Fahrenheit. They are durable enough to rarely need replacement under normal street-driving conditions.

HISTORICAL TIDBIT

Although carbon ceramic rotors have been used exclusively in Formula 1, more affordable steel rotors remain standard in most types of production car-based endurance racing, including the grueling 24 Hours of Le Mans.

KEY PERSON ·

Seeking a competitive advantage in Formula 1, the Brabham racing team's technical director Gordon Murray introduced carbon ceramic stoppers in 1979 using the byproducts from aircraft brake manufacturing. Murray went on to design the revolutionary McLaren F1.

Slowing down is the less glamorous counterpart to going fast. And while brakes have been a crucial component of automobiles from day one, their stopping power has not always kept pace with acceleration.

Early brakes had meager stopping ability. The prevalence of relatively weak drum brakes persisted for decades, a spell broken in the racing world when Jaguar adopted disc brakes for the 1953 24 Hours of Le Mans.

Disc brakes with steel rotors became the typical arrangement on high-end cars in the 1950s, and the trend lasted for decades. Innovations made in aviation—in this instance, military aviation—eventually trickled down to commercial airliners, with standard carbon brake rotors first appearing on the Concorde in 1976 and carbon-ceramic variants later becoming employed on the French high-speed TGV train.

Carbon ceramic brakes (CCBs) work using the same principal as steel brakes but utilize a blend of ceramic and carbon fiber strands to enhance heat dissipation. With better cooling comes the reduction of brake fade, which occurs when brakes lose their effectiveness following hard or repeated stops. CCBs can also weigh as little as half as much as steel discs, a crucial benefit in vehicle dynamics because they are part of a vehicle's unsprung mass (components including suspension and wheels). The lower weight of the braking hardware reduces rotational inertia and also aids handling.

Despite their benefits, it wasn't until 1979 that CCBs made their F1 racing debut with the Brabham team. Carbon ceramics finally trickled down to street cars with Porsche's 2001 911 GT2. Early models offered vague pedal feel and didn't quite deliver on the promise of infinite durability, but the technology evolved rapidly into a much more usable form.

Nearly every modern supercar is now equipped with carbon ceramic brakes, though price remains a challenge: while some brands, such as Ferrari, feature carbon ceramics as standard equipment, others charge $8,000 to $15,000 for carbon brakes as a standalone option.

TRACTION CONTROL

Power, so the cliché goes, is nothing without control. In the realm of supercars, that can certainly be the case. But the introduction of traction control in the supercar world was a controversial, and not always welcome, advance.

Traction control (TC) is a technology that detects when power to the wheels cause, wheelspin and mitigates the loss of traction (and control) by reducing or eliminating engine power and/or selectively applying the brakes. Though TC originated in Formula 1 racing in the 1980s, the technology was banned in 1994 until it became legal once again in 2001—and was banned yet again in 2008.

Production cars were essentially free of TC until the late 1980s, and the technology appeared in supercars relatively slowly. The last holdover of note was the 2005 Ford GT, an outlier whose midengine, rear-drive, supercharged 550-horsepower setup certainly could have used electronic intervention. Why the resistance? For starters, with tolerances and technology well beyond those of standard passenger cars, adding traction control to supercars, particularly crude early systems, would have neutered their wild-child personalities. In the macho world of supercar buyers, there was also a sense of pride in owning an unfettered beast that relied entirely on driver skill to stay safely on the road.

But as TC systems became more sophisticated the technology became increasingly common in supercars. Electronic stability control, a more sophisticated evolution beyond TC, was government mandated in the United States beginning with the 2012 model year.

Today's TC systems work seamlessly to optimize acceleration and control, monitoring wheelspin hundreds of times per second and adjusting power accordingly. Though a completely pure driving experience would theoretically be devoid of electronic interventions, such as traction control, stability control, and ABS (anti-lock brakes), most modern electronics systems are advanced enough that they will only improve upon driver skill under all but the most extreme cases.

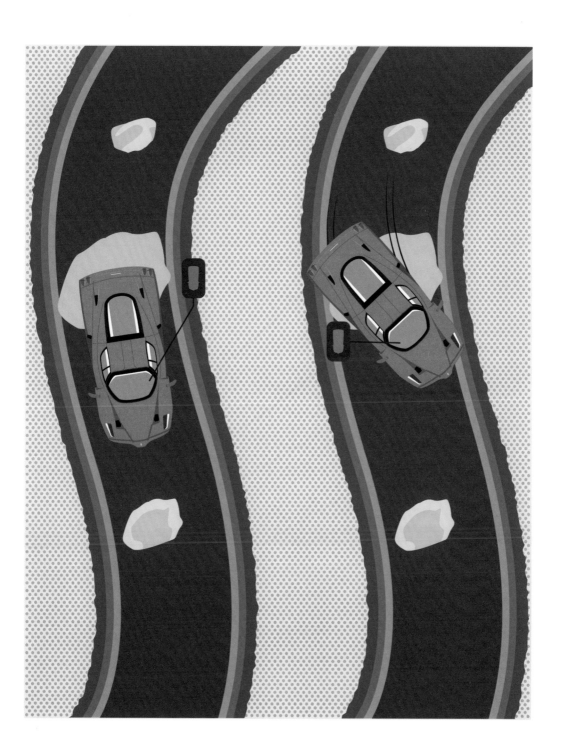

TECH REVOLUTION
ACTIVE AERODYNAMICS

Aerodynamics in relation to vehicle stability transformed automotive design, but it wasn't until the advent of active aerodynamics that engineers were able to fully maximize the power of airflow.

Striking a balance between downforce and lift is the goal of active aerodynamics, a term used to describe vehicle body components that change position in order to modify the car's aerodynamic profile—and thus, its lift/drag characteristics. As with many such features, active aerodynamics (or active aero) originated in motorsports. Following the development of so-called downforce cars in the 1960s, whose aero devices pushed the cars down with such immense force that they occasionally snapped suspension arms, wings, and other features, aero became manually adjustable. By tweaking a wing's angle of attack, a car could be configured with different amounts of downforce—more, which enabled greater grip in corners, or less, for higher top speeds.

As the nuances of airflow and drag became better understood, engineers and aerodynamicists viewed vehicle bodies as holistic entities through and over which wind traveled, taking the opportunity to incorporate specific areas of high and low pressure depending on the desired performance goal. Lotus's F1 cars of the late 1970s and early 1980s managed to reach a compromise between downforce and drag by refining ground effects aerodynamics, which controlled airflow beneath the vehicle, generating high levels of downforce without a detrimental effect on drag. Later, active aerodynamics—that is, automatically moving features that alter the vehicle's aerodynamic profiles— were introduced in racing but swiftly banned. Road cars were exempt from such rules, and thus began incorporating features that automatically altered the vehicle's aerodynamic profile.

Modern supercars are quite complex in their active aero profiles, using folding splitters, shifting ducts, and movable flaps to divert air. The most advanced of the crop can selectively shift air over the vehicle's right or left side in order to enhance cornering.

TECH REVOLUTION
HYBRID POWERTRAINS

FUN FACT

Along with the increasing complexity of hybrid powertrains comes packaging challenges: the Porsche 918 Spyder is said to be so crammed with electronics modules and high-voltage cables that a deck of cards would be impossible squeeze beneath its skin.

HISTORICAL TIDBIT

Ferrari engineers made a conscious choice when designing LaFerrari: rather than downsize the engine to a V-8, they embraced the traditional V-12—perhaps to emphasize that this car was not compromised in any way.

KEY PERSON

When it came to F1's adoption of hybrid powertrains, the buck stopped at then-FIA President Max Mosley, an ardent proponent of the technology. Mosley has since stepped down, and former boss Bernie Ecclestone has called for the scrapping of hybrids in F1.

Hybrids, which draw energy from a combination of differing powertrains, started as noble attempts to create energy-efficient, low-emissions transportation.

While the first hybrids appeared as early as 1899 (with a Ferdinand Porsche–designed model debuting in 1903), it wasn't until the Toyota Prius concept premiered in 1995 that the modern era of hybrids was ushered in.

The Prius's runaway commercial success dramatically shifted the automotive marketplace, but the performance potential of hybrid powertrains wasn't fully exploited until the so-called holy trinity of supercars emerged in 2013. The Ferrari LaFerrari, McLaren P1, and Porsche 918 Spyder turned the hybrid paradigm on its ear with blistering performance and better-than-usual efficiency considering their outrageous athletic capabilities. Though this trio's powertrains combined internal combustion and an electric motor just like the humble Prius, their mission was rooted in abject excess. The electronically controlled symphony between electric torque and screaming, gasoline-burning horsepower enabled precise handovers that enabled the seamless filling of gaps in the powerband, creating an uninterrupted torrent of power—the polar opposite of economy-minded hybrids.

This no-compromise approach to performance coincided with Formula 1's shift into hybrid powertrains, which saw small turbocharged 1.6-liter V-6 gas engines coupled with an energy recovery system allowing additional power on demand. While the hybrid F1 cars roiled fans who missed the scream of naturally aspirated V-8s, they did wonders to legitimize hybrid technology in the eyes of performance enthusiasts who might have otherwise discounted it as tree-hugging propaganda.

The hybrid trend extended with boutique manufacturers, such as Koenigsegg, and will likely continue with the next crop of high-dollar supercars.

TECH REVOLUTION
WHERE THE RUBBER MEETS THE ROAD

All the power in the world is useless if you can't lay it down to the road, and supercars, no matter how exuberantly endowed, require state-of-the-art tires to maximize their performance potential. Supercar horsepower figures have more than doubled over the past three decades—in some cases exceeding the 1,000-horsepower mark—which puts tremendous pressure on the thin rubber bands wrapped around their quickly spinning wheels.

But it's not just forward motion that stresses tires: with 200-plus-mile-per-hour top speeds, multiple g-force loading in the corners, and aircraft-carrier arrestor-cable levels of stopping power, they must cope with extreme forces in all directions while also conforming to manufacturer noise, comfort, and efficiency standards.

Although a tire's essential functionality has remained unchanged since the dawn of automotive history, meteoric rises in performance have required steep advances in sidewall construction, tread pattern, and compound formulation. While racing developments have helped advance tire technology, the challenges and compromises necessary for road-legal passenger cars are ultimately far greater than the focused needs on the track. For starters, tires within Formula 1, the top tier of auto racing, are divided into three categories: wet, dry, and intermediate. Within the dry category, seven specifications (from hypersoft to superhard) can be specified based on ambient temperatures. By honing in on three separate skill sets with further specificity within each category, race tires are able to achieve outstanding results within each area of focus.

In contrast, street tires, especially within the performance-focused supercar genre, must be capable of producing exceptional performance under varying circumstances, including weather conditions, passenger loads, and other variables. Packaging and weight constraints have also curtailed the use of spare tires, putting more pressure on tire manufacturers to engineer so-called run-flat solutions capable of keeping the vehicle safely in contact with the road in the event of a puncture or catastrophic loss of pressure.

GIAN PAOLO DALLARA

Gian Paolo Dallara is a rare impresario whose work has influenced the automotive industry from the primordial days of supercars all the way to modern times. Starting his career at Ferrari's Scuderia racing division in 1960, Dallara introduced the infamously single-minded Enzo Ferrari to the wind tunnel, where aerodynamically sound designs were incorporated into the brand's Grand Prix cars. Following a stint at Maserati where he worked on the Birdcage and Cooper-Maserati, Dallara went on to Lamborghini in 1963, where he led a team of engineers in the conception and execution of the Miura, 350 GT, and Espada.

Dallara's later work for Alejandro DeTomaso produced an innovative F2 chassis, followed by a Formula 1 project car that was campaigned by Jacky Ickx and Piers Courage with support from Frank Williams. But it wasn't until the founding of his eponymous company in 1972 that Dallara took on the racing realm as an independent force, producing a series of cars for varying levels of formula racing. Though the business had more than its share of ups and downs, including several aborted high-profile F1 projects, Dallara's investments in high-tech equipment (including several advanced race simulators and wind tunnels) has furthered his lucrative consulting business. He has offered his services in the development of supercars and race cars, including the McLaren MC12, Bugatti Veyron, Audi R8 LMP, and Lamborghini Huracan GT3. That know-how helped Dallara venture into designing and building his small-batch supercar, a project which kicked off in 2013.

Gian Paolo Dallara's work has spanned decades and influenced countless cutting-edge projects, helping push the bleeding edge of supercar design and engineering. Still going strong in an industry that is known for mercurial change and shifting approaches, Dallara has had tremendous staying power, which might have something to do with his unusual combination of forward-looking engineering and enterprising industriousness.

GLOSSARY

ANTI-LOCK BRAKES (ABS): Brakes that use modulated pulses, between fifteen and thirty times per second, to maintain control and prevent skidding during hard stopping maneuvers.

CARBON CERAMIC BRAKES: A type of brake manufactured using a resin that fuses a composite of carbon and silicon and bakes it into a diamond-like silicon carbide material. The result is a highly specialized, lightweight brake rotor that can withstand and dissipate extreme temperatures, making it ideal for ultra high-performance applications.

CARBON FIBER REINFORCED PLASTIC: CFRP, also known as carbon fiber reinforced polymer, is a strong, lightweight material that can be as much as one-fifth as light as an equally strong section of steel.

ENGINE CONFIGURATION: The layout and count of cylinders that determines an engine's packaging and proportions. Configurations include straight-eight, flat-twelves, V-12s, and others.

FLAT CONFIGURATION: An engine configuration in which the cylinders are opposed to each other in a flat configuration. Also referred to as boxer engines.

HYBRID POWERTRAIN: An engine configuration that combines more than one type of powerplant, often an internal combustion engine and an electric motor.

MONOCOQUE: A type of vehicle construction in which the body is formed as part of the chassis, forming a strong, single unit.

SIDEWALL: The section of a tire that flanks the flat contact patch on either side, connecting to the wheel via a bead.

SIPING: Thin grooves between tread blocks that enhance traction in wet, sandy, or snowy surface conditions.

STABILITY CONTROL: An electronic safety system that helps maintain vehicle stability by detecting a loss of traction and automatically adjusting engine power and/or applying individual brakes.

SUPERCHARGING: A way of boosting power by using a crankshaft-driven turbine to enhance airflow into the engine.

TRACTION CONTROL: An electronic system that detects wheel slippage and cuts engine power and/or applies brakes in order to maintain control of a vehicle.

TREAD BLOCK: Sections of rubber that form a tire's tread pattern. The spacing and shape of the tread block determines a tire's wet and dry grip characteristics.

TURBOCHARGER: A device that extracts more power from an engine by using an exhaust-driven turbine to pump a higher volume of air into the combustion chamber.

V CONFIGURATION: An engine in which the cylinders are arrayed in a V layout—usually, at 60 degrees or 90 degrees in relation to each other.

VULCANIZED RUBBER: A type of rubber that has undergone a process that combines natural rubber with an additive to make it more durable; all automotive, motorcycle, and bicycle rubber is vulcanized.

W CONFIGURATION: An engine configuration in which the cylinder banks are aligned in a shallow, or parallel, angle.

MODERN
MILESTONES

The year was 1988 and the concept seemed impossibly cool: a swoopy, V-12-powered Jaguar with scissor doors, all-wheel drive, and a model name that denoted a stunning top speed of 220 miles per hour. Fresh on the heels of the excess of the 1980s, the XJ220 seemed to have everything going for it: a mighty powerplant, drool-worthy looks, and a sense of superiority with the potential to redefine the very fabric of the Jaguar brand.

Somewhere between concept and execution, however, the British manufacturer lost its mojo as the global economy suddenly flailed. Those shifts forced crucial changes to the production version of the once-alluring concept. Gone was the V-12, swapped for a humbler twin-turbo V-6. The all-wheel-drive system was traded for a rear-drive setup, and the scissor doors were dropped in favor of conventional portals. Not surprisingly, a good portion of the hundreds of deposit holders backed out, and some even took legal action against the manufacturer for what they considered a case of bait and switch.

By the time the production XJ220 rolled out in 1992, naysayers didn't really have a lot to complain about. Sure, the scissor doors would have been novel and there *is* considerable mystique in the V-12 configuration, but the final car was also lighter and nimbler than the all-wheel-drive concept, and produced more power: 542 horsepower, versus the promised 500.

Produced until 1994, only 275 XJ220s were built—a bit less than the original production run estimate of 350. And while it was a milestone-setting performer for its time, launching to 60 miles per hour in 3.7 seconds and posting a top speed of 217 miles per hour (just shy of its promised 220-mile-per-hour terminal velocity), the XJ220 ultimately enjoyed the last laugh: not only has its swoopy design proven to stand the test of time, values of this still-attractive supercar have been trending upward, far surpassing the original MSRP.

In a world besotted with committees, compromises, and concessions, the McLaren F1 emerged in the early 1990s as a singular vision of aesthetic and functional purity. The brainchild of Gordon Murray, the F1 embraced a take-no-prisoners, spare-no-expense approach to speed. Every aspect of the F1—from its industry-first carbon fiber monocoque to its central seating position—was focused on delivering the ultimate sensory experience for the driver. And yet it wasn't just a race car bearing a license plate: the F1's cabin stayed cool thanks to real gold insulation surrounding the engine bay. Also aiding everyday usability was outstanding outward visibility and fold-out compartments that stowed fitted luggage.

Targeting the goal of supreme performance made the F1 an exercise in both mechanical excess and rigorous control. Though it weighed a mere 2,500 pounds, its naturally aspirated 6.1-liter BMW V-12 produced a considerable 627 horsepower. That fearsome power-to-weight ratio enabled a 0–60-mile-per-hour time of 3.2 seconds and a record-smashing top speed of 240.1 miles per hour in a prototype car (production models clocked closer to 231 miles per hour). Aiding downforce were two Kevlar fans, which ensured high-speed stability by literally sucking the F1 closer to tarmac. So outrageously powerful was the F1 that when the race version was built, it actually had to be detuned to be eligible for competition at the 24 Hours of Le Mans. In 1995, a GTR race model won at Le Mans in its debut attempt, no small feat for any racer, let alone one that started life as a street car.

Only 107 race and roadgoing F1s were built, making the F1 more coveted as time passes. That rarity has fueled its ever-escalating value, which most recently has climbed to the $15 million mark.

FUN FACT

Despite its state-of-the-art carbon fiber chassis, the Carrera GT's shift knob is constructed of layered birch and ash, an homage to the 917 race car's balsa wood shifter.

HISTORICAL TIDBIT

The Carrera GT's V-10 was originally destined for racing but got shelved when Porsche chose to focus its resources instead on developing the Cayenne SUV.

KEY PERSON

Racing and rally legend Walter Röhrl, who helped develop the Carrera GT, said it was "the first car in my life that I drive and I feel scared." His feedback helped convince Porsche engineers to incorporate traction control into the production version of the car.

The rear-engine configuration—that is, the practice of hanging the powerplant aft of the rear axle—is the singular hallmark of the Porsche 911, the icon responsible for much of Porsche's commercial success. But despite that stubbornly anachronistic setup, a few midengine Porsches earned their own legitimate notoriety, among them the radical Carrera GT.

Developed using mechanical vestiges from the midengine GT1 and LMP1-98 Le Mans Prototype race cars, the Carrera GT was powered by a competition-derived V-10 that was detuned for public consumption. Even in its slightly milder road legal form, the V-10 was a screamer, revving to 8,000 rpm and churning 605 horsepower. It also wasn't pushing much weight thanks to a carbon fiber monocoque that dropped the GT's weight to just over 3,000 pounds.

Compact, agile, and notoriously challenging to manage at the limit, the manual transmission-only Carrera GT was somewhat of an outlier compared to contemporaries such as the Ferrari Enzo and Mercedes SLR McLaren. Unlike its techy competitors, the Carrera GT embraced a more analog, reductionist approach that required more technique to drive fast. Gentle acceleration from a standstill demanded a careful approach, even for seasoned three-pedal enthusiasts: rather than feeding the accelerator, the ceramic clutch was designed to engage with a gentle release and no application of throttle. Hard launches were even more challenging since the engine's lightweight flywheel sped up and slowed down quickly, putting the onus on the driver to maintain a strictly disciplined use of the gas pedal.

With no stability control system in place, the Carrera GT's only electronic interventions were anti-lock brakes and traction control. Along with its tricky handling characteristics, the GT, for better or worse, earned a reputation as a widowmaker, a notoriety that became reinforced when actor Paul Walker perished in a Carrera GT that lost control on a Southern California street.

MODERN MILESTONES
FORD GT

The 1960s-era Ford GT (often referred to as GT40 for its 40.5-inch height) was an actual race car with only a handful homologated for road duty. But the twenty-first century has seen two reiterations of the race-ready classic brought to life: a version produced from 2004 to 2006 that paid homage to the Le Mans–winning era, and a 2017 redux whose racing counterpart won its class at the 2016 24 Hours of Le Mans, the fiftieth anniversary of the original GT's win.

While the two modern cars took similarly unconventional approaches to styling, their functionality and commitment to performance differed dramatically. The first-generation car was styled by Camilo Pardo and bore a visual resemblance to the 1960s-era road and race cars. Though undoubtedly capable with its supercharged, 550-horsepower 5.4-liter supercharged V-8 and innovative aluminum spaceframe chassis, the first modern GT was built as a tribute to Ford's centennial, making it more of a high-speed street car with aspirations to competition than a race car cloaked in a street car's body. Interestingly, it was among the last supercars to lack traction control, which, not surprisingly, led to more than a few of these torquey two-seaters swapping ends on public roads.

The third iteration was literally born from racing: following Ford's last-minute decision to campaign an all-new car at the 2016 24 Hours of Le Mans, the fiftieth anniversary of the manufacturer's historic 1-2-3 finish, a skunkworks team fast tracked an ultra high-performance vehicle that could satisfy homologation requirements and work both as a racer and a road car. The result was a carbon chassis, 647-horsepower beast motivated by a twin-turbo V-6 and managed by an advanced, adaptive pushrod suspension. The newest Ford GT was not only a fearsome street machine with a top speed of 216 miles per hour, its racing iteration proved to have the right stuff, finishing first, third, and fourth in its class at Le Mans on the fiftieth anniversary of Ford's original victory.

MODERN MILESTONES

FERRARI ENZO

It takes a special vehicle to sit atop the Ferrari lineup, and a truly monumental one to inherit the company founder's name. The Ferrari Enzo arrived in 2002 with jagged looks and a broad, imposing stance that a delivered a sharper visual impact than its predecessor, the swoopy F50. It was a look guided more by the wind tunnel than the whimsies of the styling house responsible for the design, Pininfarina.

Pundits had expected it to be named F60, but the Enzo's name revealed that the new package was more than just an F50 sequel. Single-mindedly focused on performance, the Enzo featured the counterintuitive combination of a carbon fiber chassis, trick outwardly opening doors, and stripped-down Formula 1-derived aerodynamics, but the lack of modern amenities such as power windows. The Enzo was the last Ferrari flagship to be powered by a naturally aspirated engine without the assistance of a hybrid system. The all-new engine was a 6.0-liter V-12 that screamed its way to an 8,200-rpm rev limiter, producing 650 horsepower and 485 lb-ft of torque. More significantly, the powerplant was mated to an automated six-speed transmission actuated by paddle shifters, the then-nascent technology that would make its way into every Ferrari and, eventually, every supercar.

The aggressive aerodynamics came from front flaps, a rear spoiler, and two rear diffusers that produced as much as 1,709 pounds of downforce at 186 miles per hour. That downforce tapered off to 1,290 pounds by the time it reached its top speed of 221 miles per hour. Reversing that speed were then-revolutionary carbon fiber brakes that had only previously been used on race cars. The unfortunate reality with the Enzo (along with virtually every other ultra-limited production supercar) is that despite the extensive race-derived engineering, climbing values have relegated most examples to languish undriven in the garages of collectors.

FUN FACT

Of the initial 400 Enzos built, one was earmarked for Pope John Paul II, who later auctioned it with the funds of just under $1 million dollars going as relief for 2004's southeast Asian tsunami. The same car was auctioned again in 2015 for $6 million.

HISTORICAL TIDBIT

Drawing from Enzo DNA was the FXX, which spun off as a track variant. Maserati's MC12 followed with twenty-five units, which were built to satisfy homologation rules.

KEY PERSON

The Ferrari Enzo was developed with assistance from multiple world championship-winning driver Michael Schumacher, whose recommendations included lowering the level of traction control intervention.

MODERN MILESTONES
LAMBORGHINI AVENTADOR

FUN FACT

While the Aventador is inarguably powered by a V-12, its powerplant doesn't always run on all twelve cylinders; when it's not under load, a cylinder deactivation system imperceptibly switches off half the cylinders to boost fuel efficiency, transforming the V-12 into an inline-six.

HISTORICAL TIDBIT

The Aventador is only the fifth wedge-shaped Lamborghini flagship in history, preceded by the Miura, Countach, Diablo, and Murcielago.

KEY PERSON

While the Aventador was penned by designer Filippo Perini, its radical mechanical underpinnings were spearheaded by engineering and R&D boss Maurizio Reggiani, who took on the arguably more crucial task of propelling the Italian brand into the twenty-first century.

An all-new V-12-powered Lamborghini flagship doesn't come along every day, and with a lineage that includes the gorgeous Miura and iconic Countach, the prism of that supercar history made the Aventador a particularly daunting challenge.

The Aventador debuted in 2011 as a radical leap for Lamborghini, which was now under ownership by Volkswagen's Audi group. Adhering to the brand's signature wedge philosophy, but incorporating a slew of advanced hardware, the Aventador no longer trailed the supercar pack, it led. Carbon fiber chassis? Check. Race-inspired inboard suspension? Absolutely. Innovative automated single-clutch transmission? Why not?

A closer look revealed some elements that worked better than others. While the independent shift rod (ISR) transmission saved precious pounds, it proved jerky and violent. The pushrod suspension was novel, but it did not mask the Aventador's bulk. At least the massive, naturally aspirated 6.5-liter V-12 was a thing of visceral beauty, bellowing its way to 60 miles per hour in under three seconds and to a 217-mile-per-hour top speed.

Limited-production models, such as the SV, boosted the Aventador's street cred and ultra-exclusive spinoffs, such as the Centenario (twenty coupes and twenty roadsters) and Veneno (only three units built total), enhanced the model's specialness. It wasn't until 2017's Aventador S that its essential architecture saw significant upgrades that addressed some of the earlier criticisms. By adding four-wheel steering, the Aventador delivered enhanced low-speed maneuverability and high-speed stability; improvements to the chassis made the whole package flow better and become more of a driver's car.

While its driving dynamics have seen significant improvements, road presence might arguably be what ensures the Aventador's place in the supercar hall of fame. Menacing, pointy, and appropriately badass, the Aventador furthers the legend of Lamborghini's wedge.

FUN FACT

The original Veyron concept featured an eighteen-cylinder engine, while the production car was downsized to only sixteen.

HISTORICAL TIDBIT

The Veyron was named after Pierre Veyron, the French driver who was hired by Bugatti as a test driver and later won the 24 Hours of Le Mans for the manufacturer in 1939. The Chiron earned its moniker after Louis Alexandre Chiron, a prolifically successful driver who raced for Bugatti, Alfa Romeo, Mercedes-Benz, and Maserati, among others.

KEY PERSON

Ferdinand Piëch was the modern-day savior of the Bugatti brand, but the originating force was Ettore Bugatti, the Italian-born designer and founder of Bugatti. Bugatti's motto was "Nothing is too beautiful and nothing is too expensive."

Volkswagen's Bugatti brand has targeted nothing less than the ultimate extremes in supercars. Resuscitating the French nameplate in 1998 was the brainchild of VW boss Ferdinand Piëch. The famously imperious leader pressed his team with three objectives at the project's launch: an excess of 1,000 horsepower, 0–60 miles per hour in under three seconds, and sufficient elegance to not embarrass its occupants when arriving at the opera. The result was the Veyron, a boundary-busting, 253-mile-per-hour, seven-figure hypercar whose technological advances made it unlike anything that preceded it. Unfortunately, its debut coincided with a global recession. The brand survived, though it took a decade to sell all 450 Veyrons and open the door to a sequel.

The Veyron's follow-up materialized in 2017. The second time around the performance escalated further: more power (1,500 horsepower versus 1,001), a higher top speed (261 miles per hour, with room left on the table for future variants to uncork), and a more graceful silhouette that made the Veyron look downright chunky. Though Piëch was no longer the boss, the Chiron's updated looks promised even more dramatic appearances at the opera.

Enabling the Chiron's hyper performance is a quad-turbocharged sixteen-cylinder engine with 95 percent new parts compared to the Veyron. Thanks to 68 percent larger turbos that operate sequentially, a flat torque curve commences at only 2,000 rpm and doesn't diminish until 6,000 rpm, offering a tugboat-like 1,180 lb-ft of torque. The result is immediate acceleration that seems to propel the Chiron through time and space at warp speed—the arc-shaped two-seater can hit 200 miles per hour faster than it takes some cars to break the speed limit. Carbon ceramic brakes, working in conjunction with the rear spoiler deployed as an air brake, haul it all down.

As supercars go, few if any cars, past or present, exude the sheer excess of the Bugatti Chiron. While simpler, louder, less refined rides with a more obvious impact may resonate with purists, the Chiron will long stand as a monument to the marriage of engineering and brawn.

PAGANI HUAYRA

FUN FACT

Pagani models are usually obscurely named and difficult to pronounce; the Huayra (pronounced why-rah) was named after the Andean god of wind.

HISTORICAL TIDBIT

Though Horacio Pagani says he loves his brand's creations like children, he has an open affection for other brands. Among his personal collection is a Porsche 918 Spyder and a Ferrari F12 TDF.

KEY PERSON

Horatio Pagani's vision for his eponymous firm came during his time working as an engineer at Lamborghini, where he failed to convince company brass of the benefits of carbon fiber, which he employed heavily with his creations.

Horacio Pagani, after defecting from Lamborghini, emerged as a supercar builder with a rare combination of aesthetic discernment and engineering prowess. His debut car, the Zonda, wowed enthusiasts and startled mainstream competitors with its exquisite design and jaw-dropping brawn.

With the supercar pump primed, Pagani struck even harder with his sophomore effort, the Huayra. The shockingly styled outlier was a visual menagerie of complex curves and finely finished details wrapped around a unique carbon and titanium tub that frames a six-liter, 700-horsepower twin-turbo V-12. The engine, sourced from Mercedes-Benz's AMG performance division, offered the reassurance of mechanical reliability, while the Huayra's styling recalled some of the most outlandish cues this side of an H. R. Giger illustration.

Huayra packed a startling amount of technology beneath its carbon fiber skin. Though the engine mated to a relatively crude automated single clutch transmission, an array of imaginative features distinguished it from higher volume exotics. For starters, the suspension utilized a fully adjustable inboard setup also featured in race cars. Making its supercar debut was a novel active aerodynamic package that used computer algorithms to actuate four flaps. Depending on the vehicle's dynamics, the flaps move in order to optimize downforce and aid cornering stability. Adding to the car's mechanical mystique are origami-like doors and clamshell panels that articulate with delicately sculpted hardware. Despite its sculptural refinements, the Huayra delivers a surprisingly raw driving experience, with a cacophony of exhaust burbles, wastegate whirs, and drivetrain drones suggesting a mechanical character that plays foil to its smooth visual features.

Though small-batch manufacturers such as Pagani may never compete with the founding fathers of the supercar scene in terms of volume production, their imaginative engineering and wholly original designs continue to satisfy the world's exotic car enthusiasts that march to the beat of their own drums.

MODERN MILESTONES
KOENIGSEGG AGERA RS

Of the very rare outsider supercar manufacturers who have carved out a viable niche for themselves, Koenigsegg managed to distinguish themselves through novel engineering and idiosyncratic design.

First introduced in 2011, the Agera model featured a twin-turbocharged V-8 producing 927 horsepower. The 1,160-horsepower Agera RS followed up as a track-focused special, offering aero features including underbody flaps and an active wing that could produce nearly 1,000 pounds of downforce. A 1-megawatt option increased engine power to a mind-boggling 1,341 horsepower.

Whereas Horacio Pagani's creations appear to be based on a careful consideration of aesthetic balance with the benefit of unique engineering solutions, Koenigsegg works the other way around, beginning with outright performance and power-to-weight ratios that are achieved through clever engineering. Koenigsegg is also distinguished by the fact that their engines are developed in-house, as opposed to Pagani, who sources their powerplants from Mercedes-AMG.

Christian von Koenigsegg's apparent obsession with numerical milestones has led to numerous record-setting achievements, among them acceleration and top speed figures that have beat big company players, such as Bugatti. But the (arguably) more magnetic charm of cars such as the Agera RS is the intricacy of their engineering and the boldness of their mission. Features such as their dihedral syncrohelix door actuation system, which uses a balletic gliding, sliding motion to remotely open and shut the doors, harken to the early age of supercars when a sense of mechanical wonderment was applied to the unlikeliest of places. But perhaps most compelling for Koenigsegg customers is the extreme rarity of their custom-made creations: as with the vast majority of Koenigseggs, Agera RSs were built to spec for twenty-five customers, making them among the scarcest supercars on the planet.

FERRARI LAFERRARI

Not only is the founder's name incorporated into this flagship Ferrari's nomenclature, it is labeled as *the* Ferrari, suggesting a quintessence to the flagship that challenged the McLaren P1 and Porsche 918 Spyder.

While the British and Germans used V-8 engines in their hybrid power-trains, Ferrari went full-bore with a naturally aspirated 6.3-liter V-12. The sonorous engine worked in conjunction with an electric motor and F1-style KERS (kinetic energy recovery system), which regenerated energy lost in brake heat. With 789 horsepower from the engine and 161 horsepower from an electric motor, the LaFerrari produced a total of 950 horsepower.

The $1.4-million LaFerrari looked and felt different than any Ferrari before and for good reason: with old man Enzo gone, there was no debating that the brand needed to embrace change in its most radical form. As such, this race-inspired road car took bold chances on a future that involved a different design language and mechanical methodology. Although Ferrari was, in a way, sticking to their naturally aspirated soul by avoiding the use of turbochargers, the hybridized powerplant was still a dramatic departure from how things used to be done. But the performance was undeniable: with 0–60-mile-per-hour acceleration in the mid-two-second range and a top speed of 218 miles per hour, the LaFerrari broke new ground for the brand and helped push the brand into a new stratosphere of technological achievement.

A total of 500 LaFerraris were sold, the final car being auctioned for $7 million, which made it the most valuable contemporary car sold at auction. Fewer Aperta open-air models were built, with one breaking the auction record again with the final example commanding $10 million.

MODERN MILESTONES
PORSCHE 918 SPYDER

The big idea behind Porsche's 918 Spyder (as with its contemporaries, the Ferrari LaFerrari and McLaren P1) was the combination of good old-fashioned internal combustion and hybrid electric power. Unlike its peers, however, Porsche incorporated a sophisticated all-wheel-drive system that directed power to all four wheels. By leveraging the instant torque from electric motors, the 918 could accelerate and slow down individual wheels, creating a torque vectoring effect that helped it seemingly defy the laws of physics by gliding through corners with preternatural ease. Also aiding low-speed agility and high-speed stability was a rear axle steering system, as well as a symphony of no fewer than fifty electronic control units orchestrating various dynamic functions.

The 918's extreme athleticism made it equally exceptional on the racetrack, most notably at the legendary Nürburgring Nordschleife, where it clocked a record-breaking lap time of six minutes, fifty-seven seconds. But the milestone wasn't without its controversies: purists claimed the 918's all-wheel-drive system was a form of cheating, since it distributed power to all four wheels and made it far easier to drive fast. Some of Porsche's notoriously orthodox fanatics were also suspicious of hybrid technology and the car's electric steering system, insisting an element of purity was removed from the driving experience.

Regardless of the early critics, history has been rather kind to the 918, particularly because of how its advanced electronics have trickled into a fair number of Porsche's high-performance hybrid street cars. And though its $845,000 price seemed steep when the cars were sold new, values have more than doubled just a few years since, making the 918 Spyder a surefire future collectible.

MODERN MILESTONES
MCLAREN P1

It took thirteen years for McLaren Automotive division to return to road car building following the brand's first venture, the monumental F1. But as crucial as the 2011's MP4-12C debut was to the brand's twenty-first-century existence, the revival wasn't truly complete until McLaren presented a properly over-the-top hypercar.

The arms race to create halo product supercars had escalated by 2013, with established players Ferrari and Porsche on the verge of releasing exclusive offerings at stratospheric seven-figure price points. While the Italians and Germans produced the LaFerrari and 918 Spyder, McLaren's take on the ultra high-performance hybrid was the bulbous P1. Lurking beneath its curved and slatted bodywork was a 3.8-liter twin-turbo V-8 from McLaren's shared platform producing 727 horsepower, working in tandem with an electric motor for a total output of 903 horsepower. The cab-forward cockpit, framed by a carbon fiber monocoque, was positioned with a prime view of the road ahead.

Active aerodynamic manipulation was a key component of the P1's formula for speed; the large rear wing was capable of producing as much as 1,322 pounds of downforce or using a Formula 1–style drag reduction system to slip through the air more easily. An advanced hydro-pneumatic suspension system was capable of tripling the stiffness of the spring rates while also altering ride height and roll/pitch control.

With the total run of 375 models sold out, McLaren went on to build 58 track-only P1 GTRs, and even fewer special order P1 LM competition cars. As for the historical significance of McLaren's entry into this rarified air, the P1 marks a return to the levels of innovation that marked the brand as a special breed within the microcosm of highly specialized supercars. Though it is unlikely to earn the long-term notoriety of the mighty F1, the P1 was a significant achievement that proved McLaren can build a worthy competitor to the best from Ferrari and Porsche.

ENZO FERRARI

By the time he passed away in 1988 at age ninety, Enzo Ferrari's life had accumulated enough drama to fill several operas, leaving a trail of intrigue that became the stuff of automotive legend.

Born in Modena, Italy, in 1898, Enzo Anselmo Ferrari had an appetite for speed, triggered when he witnessed his first race in 1908. Still, he didn't dive into that world until he was twenty and working for Construzioni Meccaniche National, where he wrenched on race cars and eventually took the wheel. He later joined Alfa Romeo, where he formed his own racing team, Scuderia Ferrari. So great was his love of racing that he only started building road cars in 1947 at the urging of his accountants.

Ferrari's early road cars earned notoriety because they seemed to share their tempestuous nature with the brand's race cars (not to mention their maker). Prime example: After Ferruccio Lamborghini, a successful tractor manufacturer from neighboring Sant'Agata, had bought several Ferraris, he grew so frustrated by their failing clutches and Enzo's disdainful treatment that he started his own car company.

Enzo's street cars may have started life to fund his racing efforts, but the symbiotic relationship between the racetrack and the road became a crucial differentiator for the brand. Despite some of the obvious technology trickle-down from race cars, Enzo still often resisted change. He was so convinced of the supremacy of front-engine/non-independent suspension designs that it wasn't until most of his competitors had adopted the superior midengine/independent construction that he finally capitulated. He also proclaimed that "aerodynamics are for people who can't build engines," a belief he eventually rescinded.

Enzo also famously went to war with Henry Ford II in 1963 following negotiations to sell his company to the American brand. After reading a contract clause that would have signed away the motorsports division, Enzo furiously balked at the deal, triggering Ford to retaliate by building the Ford GT, which eventually beat Ferrari (and the world) at Le Mans in 1966.

Enzo's death in 1988 triggered speculation that the company might fold, but it was resurrected by Luca di Montezemolo, who took the reins of Ferrari from 1991 until his resignation in 2014. Il Commendatore's inimitable spirit lives on in a generation of race and road cars that were born during his forty-year reign.

GLOSSARY

AUTOMATED GEARBOX: A manually operating gearbox that has been fitted with a system to change gears without the need for physically moving a gearshift lever.

DIHEDRAL DOORS: A more mechanically complex version of scissor doors in which the door structure moves outwards and away from the body.

KEVLAR: A synthetic fiber used in aeronautic and high-end automotive applications for its strength and heat resistance.

PADDLE SHIFTERS: Flat tabs positioned behind the steering wheel used to trigger gear-shifts; the left paddle triggers downshifts, and the right paddle triggers upshifts.

SCISSOR DOORS: Upwardly opening doors that came to be one of the defining features of the supercar genre.

SPACEFRAME CHASSIS: A body structure in which a core structure, not the outlying surfaces, are for load bearing.

SACRED GROUND

SACRED GROUND
MODENA

FUN FACT

Despite its industrious roots, Modena has become a haven for exotic car enthusiasts seeking a glimpse behind the supercar curtain, which has led the region's cottage industry of supercar rental businesses to flourish.

HISTORICAL TIDBIT

Modena is known for countless automotive claims to fame, but the region is also home to a twelfth-century cathedral and Ghirlandina tower. These cultural influences helped earn it status as a UNESCO World Heritage Site.

KEY PERSON

Titans of the supercar industry rule the Modenese area, but the late local businessman Umberto Panini stands out for his extensive collection of Maseratis and motorcycles, which manages to attract crowds to an idyllic farmhouse just outside of town.

It's no secret the land of Puccini and pasta has been a historical hotbed for supercar culture. But one particular region in Italy has played host to a statistically unlikely array of manufacturers known for outlandish automotive creations. Modena, a municipality within Italy's northern region of Emilia-Romagna, is home to some of the nation's most legendary supercar movers and shakers.

Perhaps most famously, Enzo Ferrari was born in Modena in 1898 and eventually laid roots there that would attract some of the world's top supercar brands. Young Enzo's most formative memories were forged in the area; for instance, he attended his first race in nearby Bologna at the age of ten, and impressions from that afternoon later inspired him to enter the world of motorsports. Though he moved to Milan in 1919 to work for Costruzioni Meccaniche Nazionali, he returned to Modena in 1929 to found his racing firm, Scuderia Ferrari, from which his road car company evolved.

When the Italian government's forced decentralization of businesses came about during World War II, Enzo relocated his manufacturing plant to nearby Maranello, affording him more space to expand the footprint of his business and build the famous Fiorano circuit, a 1.9-mile-long track used to develop the company's Formula 1 and road cars.

In the intervening years, Modena became populated by a number of other brands; Maserati moved from Bologna to Modena in 1937; De Tomaso was formed there in 1959; Dallara set up shop in neighboring Varano de Melegari in 1972; and Pagani was founded in nearby San Cesario sul Panaro in 1992. Along with those nameplates came countless suppliers who settled in nearby to provide many of the cutting-edge components essential to supercar construction.

Though its agrarian roots remain in the surrounding countryside, the region remains the epicenter for supercar production, earning nicknames such as "Motor Valley."

SACRED GROUND
SANT'AGATA BOLOGNESE

A short thirty-minute drive from Modena is Sant'Agata Bolognese, a tiny rural suburb of Bologna that was put on the map by Ferruccio Lamborghini. Born in nearby Renazzo, Lamborghini found success in the tractor business, which enabled him the means to purchase several Ferrari 250s. Those gave him enough grief to require constant jaunts back and forth to Ferrari's headquarters in Modena. When he finally tired of Enzo Ferrari's attitude and gave up on his own modifications to the prancing horse cars, Lamborghini formed Automobili Lamborghini in 1963.

Inspired by the brutal beauty of Spanish fighting bulls, Lamborghini's beastly logo is a graphic reflection of the manufacturer's rebellious nature. While Ferrari was the world's supercar maker of record, Lamborghini became the challenger brand bold enough to deliver an edgier, wilder style. If Ferrari was the establishment that attracted other supercar brands to set up shop in the area, Lamborghini was the bad boy from across the way, pitching an alternate take on blindingly quick, high-dollar sleds.

As Lamborghini's business grew so did its footprint on the countryside, expanding its car-making facilities to meet the global demand for its products. Ownership under the Volkswagen Group's Audi division in 1998 saw significant renovations, as well as the addition of an Advanced Composites Research Center to the campus. Building the center led to greater amounts of carbon fiber being used in the brand's road cars.

The production of Lamborghini's first SUV in decades, the Urus, promises to double the manufacturing output of the Sant'Agata plant. Though Lamborghini's factory now employs some 1,200 workers, Urus should help boost that number by nearly 50 percent. In spite of the company's internal growth, however, Sant'Agata will likely remain a predominantly agricultural outpost that retains a scrappy spirit when compared to neighboring Modena and Maranello.

SACRED GROUND
STUTTGART

FUN FACT

The 1,000-plus-horsepower Project One supercar debuted at the 2017 Frankfurt Motor Show being driven onto the stage by AMG's F1 driver Lewis Hamilton.

HISTORICAL TIDBIT

Though AMG started as a hardcore performance brand, the badge eventually trickled down to all manner of Mercedes-Benz vehicles—from small roadsters to large SUVs. To keep the brand's performance focus, AMG lately has focused on developing a hypercar, dubbed the Project One. It would transfer Formula 1 technology into a limited production road car.

KEY PERSON

Hans Werner Aufrecht and Manfred Schiek formed AMG, but it was Daimler chairman Dieter Zetsche who was responsible for making the tuning shop an in-house component of Mercedes-Benz.

Germans may be known for their pragmatic engineering solutions and reliable lock on the luxury car market, but a mean streak in the Teutonic DNA helped make Stuttgart occasionally as relevant as Italy's supercar-dense Emilia-Romagna region.

The earliest German supercar was arguably the Mercedes-Benz 300 SL Gullwing, a Stuttgart-built halo car whose racing-derived DNA defied the world's expectations of what a road car could achieve. While the 300 SL proved that a mainstream German manufacturer was capable of creating a radically aggressive sports car, Mercedes-Benz officially withdrew from motorsports following the horrific 1955 24 Hours of Le Mans endurance race in which eighty-three spectators and one driver were killed.

The unplanned side effect of that departure was that two Daimler engineers, Hans Werner Aufrecht and Manfred Schiek, left the manufacturer to form a company of their own using a combination of their initials and the 'G' from Grossaspach, the town just thirty minutes outside of Stuttgart where they set up their small workshop. Their engine expertise proved deep enough help to claim no fewer than ten German Touring Car race victories in 1965. After forming the Aufrecht Melcher Grossaspach engineering firm—AMG—in 1967, they used their prowess for building fast, reliable machines to make the company one that Mercedes, in 1999, took a majority stake in and then fully acquired in 2005.

AMG's greatest hits include the monstrous *Red Pig*, a 6.8-liter V-8-powered sedan that scored a class-winning, second overall finish at the 24 Hours of Spa; the 1980s-era Hammer based on the E-Class coupe; the SLS, a modern reinterpretation of the 300 SL; and the GT S, the brand's latest halo car.

While Germany's southern region of Bavaria has been responsible for world-renowned tuning brands, such as Alpina and Ruf, it's Stuttgart—and more specifically, the neighboring town of Affalterbach (where AMG is based), where Germany continues to prove that supercars can indeed hail from large-scale carmakers.

SACRED GROUND
NARDO

On July 1, 1975, a proving ground owned by Fiat near the coastal town of Nardo, Italy, opened its doors. Though the so-called Società Autopiste Sperimentale Nardò offered more than twenty test tracks for chassis and suspension development, it was the facility's massive 7.8-mile-long "ring" that earned the location favor among automakers—and more specifically, supercar manufacturers interested in testing and developing the upper limits of their car's performance envelopes.

With a massive 52-foot-wide surface and banked construction that enables a "neutral" speed (i.e., not having to turn the steering wheel) of 149 miles per hour, the Nardo Ring is among the world's fastest test circuits and one of the only closed tracks where a carmaker can safely test a supercar's handling traits at or near top speed. Its open spaces have seen everything from six-wheeled Formula 1 cars to experimental cars and endurance record-breakers.

Unlike handling courses that measure traits such as a car's transitional response when cornering, a banked oval is designed for high-speed and top-speed runs, which put unique strains on a vehicle's engine and suspension components. By testing a car's behavior at triple-digit speeds (which often surpass 200 miles per hour in a supercar), engineers are able to document variables such as aerodynamic drag and downforce, engine cooling, and directional stability. Things can quickly go south at these speeds, and mishaps such as blown tires or engine failures can have dramatic effects that make a closed circuit preferable to other areas where some supercars are tested.

Porsche has owned the circuit since 2012 and sometimes rents to other automakers. While other test tracks also offer high-speed validation, few, if any, possess the sheer scale of Nardo's facility. However, one of the downsides of a banked oval is that beyond 149 miles per hour, a vehicle encounters tire scrub, which can reduce its top speed by up to 3 percent; for ultimate top-speed testing conditions, a car must be driven in a straight line.

SACRED GROUND
EHRA-LEISSEN

Germany's Ehra-Leissen test track has been shrouded in secrecy since its construction in 1968, and for good reason: the Volkswagen-owned complex is where the automaker develops some of its most top-secret projects long before they are introduced into the marketplace.

Built near the then-border of East Germany because it was a no-fly zone that prevented the possibility of aerial spying, the complex offers a variety of handling courses. But most notable is its high-speed track, which uses two 5.6-mile straightaways linked by banked corners at either end. Unlike Nardo's constant radius, banked configuration, which sacrifices some of a car's top speed due to tire scrub, Ehra-Leissen's lengthy 5.6-mile straightway is the only closed circuit on earth that enables true top speed testing— making it ideal for ultimate bragging rights.

Ironically, Volkswagen does not allow outside manufacturers to use the course, choosing to focus instead on its own product portfolio. But that wasn't always the case: in 1998, a McLaren F1 driven by Andy Wallace achieved a stunning 240.14-mile-per-hour average at the track. The doors shut on outside manufacturers when Volkswagen began developing its über powerful, quad-turbocharged, sixteen-cylinder Veyron. With a prototype running in 2003, the Veyron was developed into a 1,000-horsepower beast that eventually shattered the F1's record with a 253.81-mile-per-hour top speed in 2005. The 1,200-horsepower Veyron Super Sport further annihilated that figure with a 267.856-mile-per-hour average top speed in 2010.

Bugatti's production car record has since been broken by the 1,380-horsepower Koenigsegg Agera RS's two-way top speed run on a Nevada highway, which saw an average of 277.9 miles per hour. But the fight isn't over yet; future iterations of the 1,500-horsepower Bugatti Chiron, the follow-up to the Veyron, promise even higher figures.

SACRED GROUND
NÜRBURGRING NORDSCHLEIFE

FUN FACT

The "flugplatz," or "airport," section is notorious for its tendency to send vehicles airborne: a Nissan GT-R Nismo launched end-over-end during an endurance race in 2015, killing a spectator and injuring several after it flew into a grandstand.

HISTORICAL TIDBIT

Formula 1 racing abandoned the famed circuit when Niki Lauda was nearly killed there in a fiery crash in 1976.

KEY PERSON

The all-time record at the 'Ring was achieved by racer Stefan Bellof in 1980 during a qualification lap for the 1983 Nürburgring 1,000-kilometer endurance race. His lap time of 6:11.13 was achieved in a Porsche 956 race car, earning a corner to be named after him at the 10.5-mile mark, the spot where he impacted a guardrail at 160 miles per hour shortly after his record-setting lap.

The 12.9-mile Nürburgring Nordschleife climbs, dives, and banks its way through Germany's Eifel mountain range, offering a rollercoaster ride for racers and test drivers who tackle its 187 corners. Constructed between 1925 and 1927, the circuit earned the title of "Grüne Hölle," or "Green Hell," from three-time Formula 1 champion Sir Jackie Stewart, who was both mesmerized and daunted by its series of seemingly un-learnable bends.

A small group of some of the world's most elite drivers have, indeed, learned the challenging course, though it takes hundreds of laps to fully comprehend the nuanced camber shifts, elevation changes, and blind corners that define some of the trickier sections of the track. The Nordschleife earned its challenging reputation not only because of its vast length and complexity, but because its footprint creates significant temperature and weather spreads at different spots on the track; it may be sunny and dry along one stretch while snowing at another. Also adding to its lore is little to no runoff in certain spots, which elevates the fear factor for 'Ring driving thrill seekers.

The Nürburgring has become a favored spot for vehicle development testing because it combines a wide variety of surface conditions and dynamic scenarios, from the bumpy patchwork of the banked "Karussell" (carousel), which challenges the damping capability of a car's suspension, to a total of 1,000 feet in elevation changes that tax a vehicle's weight distribution, and a straight that tests high-speed stability.

Nürburgring Nordschleife lap times have become a bragging point for carmakers wishing to boast of their road car's performance prowess, which has created an ongoing war of numbers. Street-legal supercars have managed to pull off increasingly impressive lap times, the most recent of which was a 6:47.25 run achieved by the Porsche 911 GT2 RS. However, one lap time unlikely to go contested any time soon is the outright record of the Porsche 956 race car, which lapped the circuit in an amazing 6:11.13.

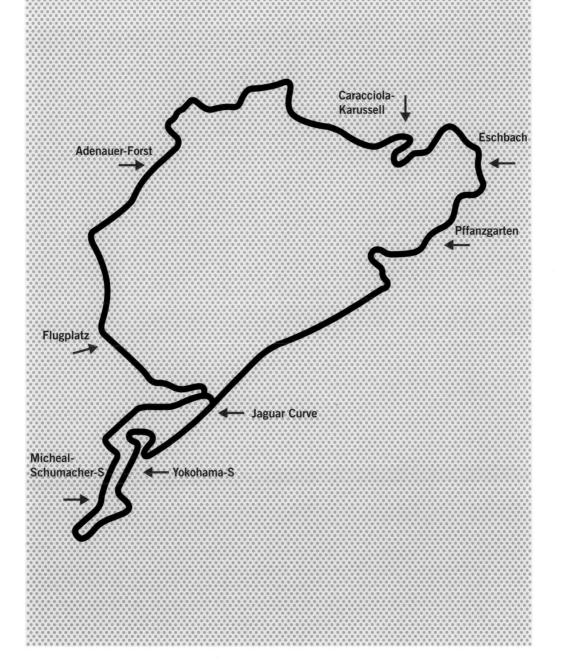

Caracciola-Karussell

Eschbach

Adenauer-Forst

Pffanzgarten

Flugplatz

Jaguar Curve

Micheal-Schumacher-S

Yokohama-S

FERDINAND PIËCH

Few automaker executives have left as indelible an impression on the industry as Ferdinand Piëch, the grandson of Porsche founder Ferdinand Porsche.

A relentlessly focused engineer and businessman, Piëch began his executive career at Porsche's racing division, where he famously spearheaded the groundbreaking 917 race car that scored the brand's first victory at the 24 Hours of Le Mans in 1970. That win established the brand's reputation as a racing powerhouse. When he left for Audi, he steered the Ingolstadt-based brand into success by pioneering the Quattro branding strategy, which established a distinct identity amidst luxury competitors such as BMW and Mercedes-Benz.

Piëch later became CEO of the Volkswagen Group in 1993 and rescued the conglomerate from bankruptcy not only through the improvement of core Volkswagen products, but also by purchasing iconic global brands, including Bentley from Britain, Bugatti from France, and Lamborghini from Italy. The transformation of those ultra-luxury brands enabled the Volkswagen Group to transcend its original vision as "The People's Car," entering the stratosphere of ultra-premium products. By revitalizing those brands from dormancy and/or extremely limited production, Piëch contributed to a modern renaissance that fueled competition among other high-end nameplates, including Fiat-owned Ferrari and BMW-owned Rolls-Royce.

The crown jewel of Piëch's tenure was arguably the complex and ambitious Veyron, which was powered by a sixteen-cylinder, quad-turbocharged engine capable of propelling it to a record-breaking top speed of 267 miles per hour. The Veyron was succeeded by the Chiron, whose future variants promise even greater top speeds.

The breathtaking arc of Piëch's career was cut short by Volkswagen's diesel scandal, which saw the colorful leader willfully resign in 2015 while pointing a finger at foe Martin Wintekorn, who also later resigned. In 2017, Piëch sold the bulk of his 14.7 percent stake in Porsche SE, the holding company that controls Volkswagen, for over a billion dollars in cash. Ruthless and brilliant, Piëch will be remembered not just for the pioneering vehicles he was responsible for but also the trail of chaos and intrigue left by his departure from the business.

GLOSSARY

APEX: The sharpest angle, or the point at the inside edge, of a corner at a racetrack. The apex is the single most crucial point that helps a driver determine his or her racing line.

TEST DRIVER: A driver employed by a carmaker to use his or her impressions to deliver feedback and guidance that helps evolve a vehicle from the prototype to production stage.

BANKING: An angled section of road or track that enables higher speeds due to the centrifugal force of the vehicle against a more vertical plane.

CAMBER: The slight tilt of a road or track surface that can boost or hurt a car's grip; off-camber surfaces lean away from the apex of a corner, forcing slower speeds, while on-camber corners enable higher speeds.

QUALIFICATION LAP: A timed lap prior to a race in which the order of the starting grid is determined. Qualifying lap times are often quicker than race laps because cars are allowed to use stickier compound tires and are unfettered from race traffic.

RACING LINE: The path of a vehicle through a corner, which can be broken down into three stages: entry, corner, and exit.

RUNOFF: The area outside of a racetrack's edges intended to allow a car to safely slide in case it loses control at speed.

TIRE SCRUB: Friction created by a tire during cornering. Ideal top speed testing scenarios require driving in a straight line, not a corner, to eliminate speed-sapping tire scrub.

V-MAX: Another term to describe terminal velocity, or top speed.

INDIE SPIRIT

INDIE SPIRIT
GUMPERT

FUN FACT

The Apollo Arrow will be built by an Italian firm MAT (Manifattura Automobili Torino), the same company who used their expertise in carbon fiber construction in collaboration with Scuderia Cameron Glickenhaus (SCG).

HISTORICAL TIDBIT

Gumpert was based on Roland Gumpert's collaboration with former Audi engineer Roland Mayer, who founded MTM (Motoren-Technik-Mayer), a performance tuning company.

KEY PERSON

Gumpert was originally founded by Roland Gumpert, who racked up no fewer than twenty-five World Rally Championship victories and four championships while he worked at Audi before he formed the supercar company.

Gumpert Sportwagenmanufaktur GmbH emerged on the supercar scene in 2004 led by Roland Gumpert, who first made his name as a director at Audi Sport. Fittingly, the upstart's debut model employed an Audi-sourced twin-turbocharged 4.2-liter V-8 that was modified to three specs, producing between 641 and 789 horsepower. Wrapped in a molybdenum spaceframe chassis and carbon fiber, the Apollo had clunky styling that polarized critics, though its track capabilities proved more than able to win them over thanks to lightweight construction and sharp dynamics. The Apollo also created enough downforce to enable it, theoretically, to drive upside at speeds above 190 miles per hour.

Unlike many independent supercars with big claims, the Apollo Gumpert actually delivered jaw-dropping performance. The Sport version posted a record lap on the Top Gear test track, and the Nürburgring Nordschleife lap record was set in 2009 with a time of 7:11.57.

The Tornante model was announced as a follow-up in 2010, offering a more luxurious and spacious interior, a 220-mile-per-hour top speed, and a sleek body from legendary Italian coachwork firm Touring Superleggera. A mockup of the car was unveiled at the Geneva Motor Show in 2011, but it never went to production. More extreme versions of the Apollo materialized in 2012, including the 780-horsepower Apollo Enraged and the 860-horsepower Apollo R, but the momentum wasn't enough to sustain the company; Gumpert filed for insolvency in 2012.

As with many indie supercar projects, Gumpert rose from the ashes again, promising a more accessible model in 2014 (dubbed the Explosion). But the company was eventually taken over by a conglomerate of Hong Kong investors in 2016, the same group that bought the De Tomaso earlier the same year. Dissociating the brand from its founder, the manufacturer was renamed Apollo Automobil GmbH and unveiled a strikingly angular, flowing design dubbed Titan. Later renamed Arrow, the V-12-powered car debuted in early 2018, promising to relaunch the latest iteration of the company.

INDIE SPIRIT
SSC

FUN FACT

Jerod Shelby started his automotive career by building Ferrari 355 replicas from Pontiac Fieros. The knowledge he gained from constructing a spaceframe for a fake Lamborghini Diablo helped him engineer the chassis for his first SSC-branded venture.

HISTORICAL TIDBIT

The Bugatti Veyron Super Sport briefly stole the top speed crown from the SSC Ultimate Aero with a 267.8-mile-per-hour record in 2010, but the title was revoked when it was learned that Bugatti had deactivated the production car's speed limiter, which returned the record to SSC.

KEY PERSON

Company founder Jerod Shelby isn't related to Carroll Shelby, but his car broke the speed record set by the elder Shelby's Ford GT some forty years earlier.

SSC (Shelby SuperCars) is the brainchild of Jerod Shelby, a medical device entrepreneur who bore no relation to Carroll Shelby. Shelby raced go karts as a kid, which triggered a lingering obsession with supercars. He eventually went down the improbable path of not only building his own street-legal production car but surpassing Bugatti's production car speed record—a real-life David-and-Goliath outcome.

SSC's first venture was the Ultimate Aero, a low-slung two-seater that combined an outrageously powerful engine with a forgettable body design. The Ultimate Aero's carbon fiber body bore more than a passing resemblance to the Lamborghini Diablo, which is not surprising given Shelby's previous exploits as a supercar replica builder. The body offered a low coefficient of aerodynamic drag and outwardly opening doors that offered a small consolation for the otherwise bland styling. More crucial to speed was the Ultimate Aero's powerplant, a pushrod-operated 6.3-liter Corvette V-8, which utilized two large turbochargers to deliver a robust 1,183 horsepower and a staggering 1,094 lb-ft of torque.

The goal of the $650,000 SSC Ultimate Aero was to shake up the supercar industry by breaking the Guinness World Record for the fastest production car, a title held at the time by the Bugatti Veyron's top recorded speed of 253.8 miles per hour. Unlike the Volkswagen Group's seven-figure hypercar, the SSC was rear-wheel drive not all-wheel drive, making it harder to lay its power down to the road. The Ultimate Aero also lacked traction and stability control systems, making it more potentially dangerous to speed seekers. On September 13, 2007, along a closed stretch of Washington highway, the Ultimate Aero achieved a record-breaking two-way average of 256.18 miles per hour, barely edging out the Bugatti's top run.

The Ultimate Aero took its place in the history books as the first American carmaker to claim a speed record since the Ford GT's title in 1967, and though SSC's follow-up model dubbed Tuatara was announced in 2011, the car remains under development.

INDIE SPIRIT
HENNESSEY

Aftermarket speed merchants just can't resist striving for seemingly unreachable numbers. John Hennessey, a Texas-based tuner, is a prime example of this innately American desire for brash style and outsized power.

Hennessey got his start racing a modified Mitsubishi 3000GT and a tuned Dodge Viper at places such as Pikes Peak and the Bonneville Salt Flats. He later went into business serving the seemingly redundant role of boosting the horsepower of already mighty V-10-powered Vipers into the quadruple digits. His twin-turbocharged Hennessey Venom 1000 sent shockwaves through the enthusiast community by scoring the top spot in *Road & Track* magazine's 2007 "Speed Kings" issue, beating established models such as the Bugatti Veyron by sprinting from 0 to 200 miles per hour in 20.3 seconds.

That spirit of conquest inspired Hennessey to take on a far more lightweight platform and endow it with otherworldly power, with the goal of setting a production car speed record. The search for a wispy donor car led the Texas firm to the Lotus Exige, whose bonded aluminum chassis contributed to a total mass of just over 2,000 pounds. The donor car's modest four-cylinder engine, however, was supplanted by a twin-turbocharged 7-liter V-8 that produced an astounding 1,244 horsepower, transforming the wide-bodied Frankenstein car into nothing short of an earthbound-guided missile. The subsequent creation was dubbed the Venom GT, and it embarked on a series of record-seeking runs between 2013 and 2014 that resulted in a Guinness book-verified 0 to 300 kilometer-per-hour (186-mile-per-hour) time of 13.63 seconds and an unofficial, one-way top speed run of 270.49 miles per hour.

Only thirteen Venom GTs were sold in total, including a "Final Edition" model for $1.25 million. While Hennessey continues his sports car and truck tuning business, he is working on his own midengine supercar, the F5. A prototype was unveiled at the SEMA show in Las Vegas in 2017.

SALEEN

FUN FACT

Saleen dabbled in aftermarket electric car tuning when it altered the Tesla Model S. Rather than meddling with the sophisticated sedan's electronics, select components were modified, such as gear ratios, brake hardware, and aerodynamics.

HISTORICAL TIDBIT

Though the S7 defined Saleen's venture into the supercar realm, the manufacturer announced a more modest vehicle at the 2017 Los Angeles Auto Show. Labeled the S1, the new car is expected to cost closer to $100,000 and features a 450-horsepower turbo-charged four-cylinder engine.

KEY PERSON

Southern California native Steve Saleen's car career began when he entered club racing, which escalated to Formula Atlantic and SCCA Trans-Am. He founded Saleen Autosport in 1983 and departed the company in 2007, only to rejoin again 2012.

Steve Saleen took a deep dive into the aftermarket game in the 1980s and 1990s, producing go-fast parts, such as superchargers and suspension components, for cars such as the Ford Mustang. But it wasn't until 2000 that Saleen took the ultimate leap by entering the supercar fray with a ground-up car of his own.

His $400,000 midengine creation was dubbed the S7, which distinguished itself visually with a plethora of scoops, louvers, and extractors accenting the carbon fiber and aluminum honeycomb-trimmed body. Saleen claimed that the S7 was the most aerodynamically efficient road car on the planet and created so much downforce it could drive upside-down at speeds above 160 miles per hour.

Powered by an aluminum, naturally aspirated 7.0-liter V-8 derived from a Ford 351 Windsor small-block engine, the S7 produced 550 horsepower and 525 lb-ft of torque. Coupled with the car's sub-2,900-pound curb weight, the S7 was capable of a claimed 0 to 60-mile-per-hour time of 2.8 seconds and a top speed of 248 miles per hour. Complementing the S7 road cars were S7-R competition models that earned legitimate race successes, including no fewer than seven GT racing championships.

In 2005, the $585,000 S7 Twin Turbo debuted with a twin-turbocharged setup derived from the S7-R race car that extracted 750 horsepower from the big, pushrod-actuated V-8. By modifying the rear spoiler and diffuser for even more aggressive aerodynamics, the S7 Twin Turbo delivered even greater downforce; unfettered by restrictions from racing regulations, the street car managed to outperform the race car in several respects.

The S7 was reborn yet again in 2017 when Saleen announced the S7 Le Mans Edition, which would be limited to seven examples commemorating the race car's competition successes. Featuring the twin-turbo 7-liter V-8 and further enhancements to aerodynamics, the latest S7 is also saddled with a definitively twenty-first-century asking price of $1 million.

INDIE SPIRIT
NOBLE

FUN FACT

Lee Noble's return to automotive normalcy was manifest with the cheekily-named Exile Bug:R, a $25,000 beach buggy-style, offroad-ready sports car that borrows heavily from the Ford Mondeo parts bin.

HISTORICAL TIDBIT

Though the Bug:R departs from the supercar formula, Lee Noble's Specialized Sporting Vehicles firm is also building the high-strung Exile sports car, which promises an astounding 522 horsepower per ton.

KEY PERSON

Lee Noble created balanced, finely tuned cars that offered outstanding feedback. Interestingly, entrepreneur Peter Dyson's takeover of the company offered a similar yet more intensely focused sensibility: high-powered engines and tightly engineered chassis without the intrusion of electronic intervention.

Though independent supercar manufacturers tend to crash the scene with a stratosphere-punching, headline-grabbing debut model, Noble Automotive Limited was founded in Leeds, England, by Lee Noble, who possessed more, ahem, noble aspirations.

The British native was steeped in the supercar world in the early 1980s when he developed cars such as the Ultima Mk1, a radically styled two-seater, which, at first glance, appeared to have more in common with a race car than a street-legal machine. Noble later sold the company in 1992 and, after collaborating with Ascari on their track-focused supercars, Lee set out on his own in 1999 by building the more terrestrial M10, which quickly evolved into the punchier M12. Though a low-volume production model, the M12 was an immediate hit, attracting attention for its finely tuned chassis and overall usability.

Utilizing a tuned Ford Duratec V-6 powerplant, the M12 had modest underpinnings but surprising performance thanks to an outstanding power-to-weight ratio; 0–60 miles per hour could be achieved in a supercar-like, low three-second range, and a top speed approached 170 miles per hour. The South Africa–built M400 was produced in 2004 and upped the ante with more potent performance and a top speed of 187 miles per hour. The model was imported to the United States by Ohio-based 1G Racing under the Rossion nameplate.

With his brand's cars becoming increasingly expensive and performance-oriented, Lee Noble departed his eponymous company in 2008 in order to focus on building more attainable cars. American entrepreneur Peter Dyson bought the company and produced an altogether more ambitious, supercar-like Noble M600 in 2009. Boasting a carbon fiber chassis, a 650-horsepower twin-turbo V-8, and a 225-mile-per-hour top speed, the M600 was conceived as a sort of spiritual successor to the Ferrari F40. Though its $290,000 price tag far eclipsed that of any Noble before it, the British creation finally hit the numerical benchmarks that qualified it as a bonafide supercar.

INDIE SPIRIT
CIZETA

FUN FACT

The V-16 engine's transverse (i.e., sideways) position contributed to the Cizeta's enormous 81-inch girth, making it one of the widest production cars in history.

HISTORICAL TIDBIT

Though Zampolli's supercars went the way of the dodo bird in the 1990s, Cizeta reappeared in the news in 2009 when US Immigration and Customs Enforcement agents seized a V16T after it overstayed its welcome in the States.

KEY PERSON

Cizeta was named after the sound of Claudio Zampolli's initials in Italian, but the company was originally named Cizeta-Moroder after musician and disco pioneer Giorgio Moroder's financial contributions to, and 50 percent stake in, the company. Prior to Moroder's partnership, film actor Sylvester Stallone was poised to be a stakeholder in the company.

In many ways, the Cizeta V16T was the Lamborghini that wasn't, a one-off whose concept and execution pushed far beyond the bounds of reason to make it an exceptionally excessive supercar. Founded by Claudio Zampolli, a Los Angeles exotic car dealer and former Lamborghini engineer, the Cizeta was based on Zampolli's decade-long ambition to build a car espousing his "more is more" philosophy.

The V16T was styled by the legendary Marcello Gandini, father of the modern wedge car, using the original design study for the Lamborghini Diablo. With four popup headlights, a broad stance, and straked air intakes on the hood, haunches, and door pillar, the V16T utilized the sharper lines that Lamborghini's then owners, the Chrysler Group, had deemed too aggressive for the production Diablo. Lamborghini engineers were responsible for the V16T's mechanicals, which featured a 6.0-liter sixteen-cylinder engine essentially consisting of two flat-plane V-8 powerplants sourced from a Lamborghini Urraco and fused together. The massive V-16 dispatched 540 horsepower and 400 lb-ft of torque through a five-speed manual transmission to the rear wheels and was capable of revving to 8,000 rpm.

Contrary to the drivetrain's supernumerary configuration, the interior gauges were much reduced, a feature Zampolli felt was essential to the purity of the driving experience. With only two gauges—a speedometer and tachometer—additional information was conveyed with warning lights that displayed in increasingly alarming colors that culminated in flashing red.

Manufactured in Modena, the Cizeta V16T premiered at the Los Angeles Auto Show in 1989. Though originally estimated to cost $300,000—roughly a third more than a Lamborghini Diablo or Ferrari 512TR—by 2002 the V16T was saddled with a dizzying $649,000 price tag. Only nine cars were delivered by the halt of production in 1995, long after co-founder Giorgio Moroder departed the company. Though a roadster concept was revealed at Monterey's Concorso Italiano event in 2003, the nine production coupes marked the end of the line for Zampolli's dream.

CHRISTIAN VON KOENIGSEGG

Supercar superheroes sometimes hail from the unlikeliest of places. In the case of Koenigsegg founder Christian von Koenigsegg, that place is the Scandinavian country of Sweden. Exhibiting a fascination with cars since a young age, Koenigsegg nurtured his passion and started his self-named car company at the age of twenty-two with the goal of building advanced, lightweight supercars. It took eight years for Koenigsegg to sell his first car, but only one more year for the Guinness Book of World Records to recognize the CC8S as the world's most powerful production car, with its modified, midmounted supercharged 4.7-liter Ford engine. Only six examples were built, but the CC8S's notoriety helped further Koenigsegg's momentum in conceptualizing technologically advanced supercars that ventured far beyond the engineering limits encountered by conventional carmakers. Using advanced materials, 3D parts printing, and novel details such as a multi-hinged door assembly using a so-called dihedral syncro-helix actuation system, Koenigsegg stood apart from bigger companies by offering cars that delivered a singular interpretation of innovation and modernity.

While Koenigsegg achieved supercar landmarks such as the One:1, which offered a landmark 1:1 horsepower-per-kilogram ratio, his most buzzworthy accomplishment to date involves an Agera RS, a lonely 12-mile stretch of closed-off Nevada highway, and an afternoon in which a red and black supercar broke five world records. The most notable among those was an average two-way speed of 277.9 miles per hour, shattering the Bugatti Veyron Super Sport's 267.8-mile-per-hour speed.

While the usage of carbon fiber, high-powered engines, and advanced features such as active aerodynamics and novel construction techniques are not exclusive to this Swedish manufacturer, Christian von Koenigsegg's risk taking and boundary pushing approaches are what separate him from other brands that take a more conservative approach to supercars. By tackling the future with blue-sky thinking and a healthy combination of science and optimism in the power of technology, Koenigsegg is helping fuel excitement in an industry that can use every ounce of competition that comes its way.

GLOSSARY

BONDED ALUMINUM CHASSIS: A style of chassis that fuses extrusions, or die-cut sections of aluminum, together using glue. This structure distributes loads more evenly and strongly than welded aluminum.

BONNEVILLE SALT FLATS: A 40 square-mile stretch of salt deposits in northwestern Utah that serves as a mecca for top-speed record-setting runs.

FLAT-CRANK: A type of engine crank that is smaller and lighter, producing crank throws at 180-degree intervals, which creates a distinctively loud exhaust note.

MOLYBDENUM: A strong metal alloy used in chassis tube construction.

PUSHROD-ACTUATED ENGINE: An engine configuration in which pushrods link to a camshaft that is not positioned overhead; this is contrary to an overhead cam engine (OHC) or dual overhead cam engine (DOHC).

STRAKES: Slats on the otherwise flat surface of an air intake. The lines often serve the purpose of preventing large objects from entering the engine, though they can also be used decoratively as well.

TWO-WAY AVERAGE SPEED: Part of the Guinness Book of World Records testing rules require a car to make a top-speed run in opposite directions within a one-hour period. The goal is to prove that mitigating factors such as wind and elevation change are balanced out during an attempt to break a top-speed record.

INDEX

Look for other titles in the
SPEED READ SERIES

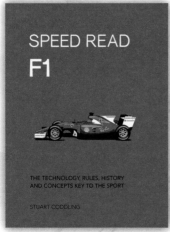

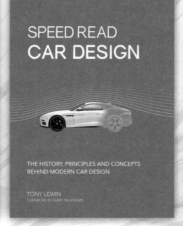

SPEED READ F1
Stuart Codling
ISBN: 9780760355626

SPEED READ CAR DESIGN
Tony Lewin
ISBN: 9780760358108

SPEED READ FERRARI
Preston Lerner
ISBN: 9780760360408

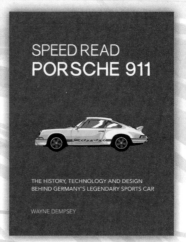

SPEED READ MUSTANG
Donald Farr
ISBN: 9780760360422

SPEED READ PORSCHE 911
Wayne Dempsey
ISBN: 9780760360422